Springer Series in Optical Sciences

More information about this series at http://www.springer.com/series/624

Pavel Polynkin • Ya Cheng
Editors

Air Lasing

Springer

Editors
Pavel Polynkin
College of Optical Sciences
University of Arizona
Tucson, AZ, USA

Ya Cheng
Shanghai Institute of Optics
and Fine Mechanics
East China Normal University
Shanghai, China

ISSN 0342-4111 ISSN 1556-1534 (electronic)
Springer Series in Optical Sciences
ISBN 978-3-319-65219-1 ISBN 978-3-319-65220-7 (eBook)
https://doi.org/10.1007/978-3-319-65220-7

Library of Congress Control Number: 2017961820

Printed on acid-free paper

This Springer imprint is published by Springer Nature
The registered company is Springer International Publishing AG
The registered company address is: Gewerbestrasse 11, 6330 Cham, Switzerland

Preface

The word laser usually refers to a complex instrument made of various optical elements, including an active medium, an optical resonator, and a power supply. In most applications of lasers, such as micromachining, the laser source is brought to the point of service where it performs its function. However, in some important applications, placing the laser at the required location is problematic. In the case of remote atmospheric sensing, such a location would be behind the region of interest, with the laser beam propagating through that region, toward the observer on the ground. Instead of using laser-like backward-propagating optical probes, the schemes of remote sensing in use today rely on the detection of the backscattering of forward-propagating lasers. The weakness of backscattering in the air negatively impacts the signal-to-noise ratios attainable with such detection schemes.

About a decade ago, it was suggested that the constituents of the air itself could be used as an active laser medium, creating a backward-propagating, impulsive, laser-like radiation emanating from a remote location in the atmosphere [1]. That is the essence of the concept of air lasing, which is the subject of this book.

The development of a viable scheme for air lasing is a challenging task. The obstacles appear to be overwhelming: The choice of the gain medium is limited as the atmosphere contains significant amounts of only nitrogen and oxygen, as well as about 1% of argon. Thus molecular, atomic, or ionic forms of these three elements will have to be used. As it turns out, molecular oxygen, being a poor candidate for the lasing medium itself, quenches lasing in the schemes that involve nitrogen emission lines. Another limitation is associated with the fact that remote air laser, by definition, has to be mirror-less. Thus optical gain needs to be very high in order to support single-pass impulsive lasing. The natural choice for energizing remote air lasing is by optical pumping with forward-propagating pump lasers. Such a traveling-wave pumping arrangement almost always favors forward-propagating stimulated emission at the expense of the emission backward. Furthermore, harmonics and broadband supercontinuum generated by the intense pump laser beam seed forward-propagating lasing, which puts the backward-propagating lasing at even more severe disadvantage as it is not seeded. Finally, remote delivery of intense pump beams is challenging due to absorption, scattering, and turbulence in the real atmosphere.

Notwithstanding the above challenges, air lasing is a subject of intense investigations by many research groups working in the field of strong-field laser physics. These efforts are motivated by both the tremendous practical potential that air lasing has in revolutionizing remote atmospheric sensing and by a purely academic challenge brought about by the complex and interdisciplinary nature of the problem. This book provides a comprehensive overview of the history and the current state of this dynamic field. The individual chapters are provided by scientists who have made significant contributions to the science of air lasing in the past; many of whom continue to actively work on this problem today.

This book begins with a chapter contributed by Joakim Bood and Marcus Aldén, who review their early work on stimulated emission in atomic oxygen and nitrogen pumped through a two-photon-resonant process. They discuss this effect from the angle of its relevance in combustion diagnostics, where, interestingly, it is sometimes viewed as an impairment. Stimulated emission lacks spatial resolution, which is necessary for imaging of combustion flows, but efficiently depopulates the upper emission state of the atom, thus quenching the useful, spatially resolving fluorescence signal.

The next chapter is contributed by Arthur Dogariu and Richard Miles. They continue the discussion of stimulated emission in atomic oxygen and nitrogen, as well as in argon, but now in the context of air lasing. Their first publication on this subject [2] spearheaded the surge of research on air lasing in the last 5 years. As of today, the lasing scheme based on the dissociation of molecular oxygen and nitrogen and subsequent pumping of atomic fragments remains the only scheme that produces lasing in the atmosphere in the backward direction, which is the most relevant for applications. The major obstacles toward the implementation of this approach at a practically relevant standoff distance are the high absorption and scattering of the deep-ultraviolet pump light that is required to pump population inversion in this scheme. We note that, counterintuitively, under a broad range of conditions, including both nanosecond and femtosecond [3] pumping, emission in the backward direction in this lasing scheme is about an order of magnitude more intense than the forward-propagating emission. The identification of the physical mechanism behind this significant forward-backward gain asymmetry is an important open problem.

The next three chapters discuss an alternative approach to air lasing, based on the emission from the nitrogen gas pumped by high-intensity femtosecond laser pulses. Two distinctly different emission channels are discussed.

In the first scheme, neutral nitrogen molecules are pumped through the collisions with the energetic electrons produced via strong-field ionization. The average electron energy needs to be above ~15 eV, in order for a predominant excitation of the upper, not the lower, emission level in the nitrogen molecule to occur. Such high electron energies are not attainable in air or pure nitrogen at atmospheric pressure when linearly polarized near-infrared laser pulses are used (e.g., the pulses naturally produced by ultrafast Ti/Sapphire laser systems). Three alternative ways of producing electrons with high-average electron energy are (i) pumping by circularly polarized near-infrared femtosecond laser pulses [4], (ii) pumping by mid-infrared femtosecond laser pulses [5], and (iii) pumping by picosecond laser pulses [6]. In the latter case, the electrons are heated through the inverse Bremsstrahlung process, involving multiple electron-molecule collisions. The mechanisms of the creation of population inversion in the above three schemes are more or less understood. However, no spontaneous lasing in the backward direction has been achieved in the natural air using either of them. The reasons for that are discussed in detail by Danill Kartashov, Mikhail Shneider, and Andrius Baltuska in Chap. 5.

In the second scheme of molecular nitrogen lasing, population inversion appears to occur "naturally" in singly ionized nitrogen molecules (N_2^+) that result from strong-field ionization of nitrogen gas. The key publication that provided important clues about the physics of the effect and stimulated follow-up investigations of this scheme is Ref. [7] by J. Yao et al. Unlike the scheme of nitrogen lasing based on pumping through electron impact, discussed above, the physical mechanism of lasing in N_2^+ remains controversial to this day. It is likely that several pumping mechanisms are operative at the same time, and the relative magnitudes of their contributions depend on the parameters of a particular experiment, the most important parameters being the wavelength and the intensity of the optical driver field. As in the scheme based on neutral nitrogen, no spontaneous backward-propagating lasing in N_2^+ has been achieved thus far. Furthermore, the presence of oxygen appears to be extremely detrimental to the N_2^+ lasing. That is likely because oxygen, having lower ionization potential than nitrogen, is predominantly ionized; electron plasma produced through ionization of molecular oxygen clamps the intensity of the pumping laser field at a lower level than what is necessary to efficiently ionize nitrogen.

Lasing in both neutral and singly ionized molecular nitrogen can be pumped by intense femtosecond laser pulses propagating in air in the self-channeling (or filamentation) regime. Laser filamentation has been shown to be resistant to the adverse propagation conditions of real atmosphere. It results in the formation of longitudinally extended regions with sustained optical intensity in the range $10^{13} - 10^{14}$ W/cm^2. Filaments can be straightforwardly produced at a standoff distance with visible, near-infrared or mid-infrared laser pulses that experience very low propagation and scattering losses in air. All of the above properties of air filamentation make the filament-based schemes of air lasing very promising. In spite of the numerous hurdles mentioned above, nitrogen air lasing remains a very active area of research where significant progress is being made continuously.

This book concludes with a chapter by Johanan Odhner and Robert Levis, who discuss remote Raman sensing in the atmosphere. This field is the primary driver for air lasing research, as it will be entirely revolutionized by taking advantage of the yet-to-be-developed remote air lasing sources. The utilization of air lasing in the established Raman sensing schemes will enable single-ended standoff sensing, where only forward-propagating excitation lasers are used, and strong and directional backward-propagating emission from air lasing is detected, resulting in high-sensitivity measurements.

P. P. acknowledges the support of his work on air lasing from the US Air Force Office of Scientific Research, most recently through the Multi-University Research Initiative on femtosecond mid-infrared laser physics, program #FA9550-16-1-0013. Y. C. acknowledges the support from the National Basic Research Program of China (Grant No. 2014CB921303) and the National Natural Science Foundation of China.

Tucson, AZ, USA Pavel Polynkin
Shanghai, China Ya Cheng

References

1. Q. Luo, W. Liu, S.L. Chin, Lasing action in air induced by ultra-fast laser filamentation, Appl. Phys. B **76**, 337 (2003)
2. A. Dogariu, J. B. Michael, M. O. Scully, and R. B. Miles, High Gain Backward Lasing in Air, Science **331**, 442 (2011)
3. S. Ališauskas, A. Baltuška, R. Boyd, P. Polynkin, Backward air lasing with femtosecond pumping, Conf. Lasers and Electrooptics (CLEO) Europe 2015, postdeadline paper PD-A-4
4. S. Mitryukovskiy, Y. Liu, P. Ding, A. Houard, A. Mysyrowicz, Backward stimulated radiation from filaments in nitrogen gas and air pumped by circularly polarized 800 nm femtosecond laser pulses, Opt. Express **22**, 12750 (2014)
5. D. Kartashov, S. Ališauskas, A. Pugžlys, M. N. Shneider, A. Baltuška, Theory of a filament initiated nitrogen laser, J. Phys. B **48**, 094016 (2015)
6. D. Kartashov, S. Ališauskas, A. Baltuška, A. Schmitt-Sody, W. Roach, P. Polynkin, Remotely pumped stimulated emission at 337 nm in atmospheric nitrogen, Phys. Rev. A **88**, 041805(R) (2013)
7. J. Yao, B. Zeng, H. Xu, G. Li, W. Chu, J. Ni, H. Zhang, S. L. Chin, Y. Cheng, Z. Xu, High-brightness switchable multiwavelength remote laser in air, Phys. Rev. A **84**, 051802(R) (2011)

Contents

1 **Diagnostic Properties of Two-Photon-Pumped Stimulated Emission in Atmospheric Species** 1
Joakim Bood and Marcus Aldén

2 **High-Gain Air Lasing by Multiphoton Pumping of Atomic Species** 19
Arthur Dogariu and Richard Miles

3 **The Role of Electron Collisions in Lasing in Neutral and Singly Ionized Molecular Nitrogen** 45
Yi Liu, Pengji Ding, Aurélien Houard, and André Mysyrowicz

4 **Molecular Rotational Effects in Free-Space N^+_2 Lasers Induced by Strong-Field Ionization** 75
Jinping Yao, Bin Zeng, Wei Chu, Haisu Zhang, Jielei Ni, Hongqiang Xie, Ziting Li, Chenrui Jing, Guihua Li, Huailaing Xu, and Ya Cheng

5 **Filament-Initiated Lasing in Neutral Molecular Nitrogen** 89
Daniil Kartashov, Mikhail N. Shneider, and Andrius Baltuska

6 **Filament-Assisted Impulsive Raman Spectroscopy** 121
Johanan H. Odhner and Robert J. Levis

Index 139

Contributors

Marcus Aldén Department of Physics, Division of Combustion Physics, Lund University, Lund, Sweden

Andrius Baltuska Photonics Institute, Vienna University of Technology, Vienna, Austria

Joakim Bood Department of Physics, Division of Combustion Physics, Lund University, Lund, Sweden

Ya Cheng Shanghai Institute of Optics and Fine Mechanics, East China Normal University, Shanghai, China

Wei Chu State Key Laboratory of High Field Laser Physics, Shanghai Institute of Optics and Fine Mechanics, Chinese Academy of Sciences, Shanghai, China

Pengji Ding Laboratorie d'Optique Appliquée, ENSTA ParisTech, CNRS, Ecole Polytechnique, Université Paris-Saclay, Palaiseau, France

School of Nuclear Science and Technology, Lanzhou University, Lanzhou, China

Arthur Dogariu Department of Mechanical and Aerospace Engineering, Princeton University, Princeton, NJ, USA

Aurélien Houard Laboratorie d'Optique Appliquée, ENSTA ParisTech, CNRS, Ecole Polytechnique, Université Paris-Saclay, Palaiseau, France

Chenrui Jing State Key Laboratory of High Field Laser Physics, Shanghai Institute of Optics and Fine Mechanics, Chinese Academy of Sciences, Shanghai, China

Daniil Kartashov Friedrich-Schiller University Jena, Jena, Germany

Robert J. Levis Center for Advanced Photonics Research, Department of Chemistry, Temple University, Philadelphia, PA, USA

Guihua Li State Key Laboratory of High Field Laser Physics, Shanghai Institute of Optics and Fine Mechanics, Chinese Academy of Sciences, Shanghai, China

Ziting Li State Key Laboratory of High Field Laser Physics, Shanghai Institute of Optics and Fine Mechanics, Chinese Academy of Sciences, Shanghai, China

Yi Liu Laboratorie d'Optique Appliquée, ENSTA ParisTech, CNRS, Ecole Polytechnique, Université Paris-Saclay, Palaiseau, France

Shanghai Key Lab for Modern Optical System, University of Shanghai for Science and Technology, Shanghai, China

Richard Miles Department of Mechanical and Aerospace Engineering, Princeton University, Princeton, NJ, USA

Andre Mysyrowicz Laboratorie d'Optique Appliquée, ENSTA ParisTech, CNRS, Ecole Polytechnique, Université Paris-Saclay, Palaiseau, France

Jielei Ni State Key Laboratory of High Field Laser Physics, Shanghai Institute of Optics and Fine Mechanics, Chinese Academy of Sciences, Shanghai, China

Johanan H. Odhner Center for Advanced Photonics Research, Department of Chemistry, Temple University, Philadelphia, PA, USA

Mikhail N. Shneider Department of Mechanical and Aerospace Engineering, Princeton University, Princeton, NJ, USA

Hongqiang Xie State Key Laboratory of High Field Laser Physics, Shanghai Institute of Optics and Fine Mechanics, Chinese Academy of Sciences, Shanghai, China

Huailaing Xu State Key Laboratory on Integrated Optoelectronics, College of Electronic Science and Engineering, Jilin University, Changchun, China

Jinping Yao State Key Laboratory of High Field Laser Physics, Shanghai Institute of Optics and Fine Mechanics, Chinese Academy of Sciences, Shanghai, China

Bin Zeng State Key Laboratory of High Field Laser Physics, Shanghai Institute of Optics and Fine Mechanics, Chinese Academy of Sciences, Shanghai, China

Haisu Zhang State Key Laboratory of High Field Laser Physics, Shanghai Institute of Optics and Fine Mechanics, Chinese Academy of Sciences, Shanghai, China

Chapter 1
Diagnostic Properties of Two-Photon-Pumped Stimulated Emission in Atmospheric Species

Joakim Bood and Marcus Aldén

1.1 Introduction

Two-photon excitation provides access to atoms and molecules with absorption resonances in the vacuum ultraviolet (VUV), i.e., to the species otherwise inaccessible for probing under atmospheric conditions, since the atmosphere strongly absorbs VUV radiation. In addition, for high enough laser intensities, two-photon pumping may create population inversion between the pumped energy state and a lower-lying intermediate state, resulting in stimulated emission. In this chapter such an emission is discussed in terms of its diagnostic capacity. The method has primarily been investigated for the detection of a number of atomic species, such as oxygen and nitrogen, and a few small molecules, for example, CO and NH_3. The major benefits of the technique are that the signal propagates in a laser-like beam, the backward-directed beam allows single-ended diagnostics, strong signals allow trace-level detection, and the optical setup is relatively simple. The main disadvantages are the poor and sometimes ambiguous spatial resolution and the difficulties with modeling the process due to the nonlinear dependence of the signal on the concentration of the active species and the integrative-growth nature of the signal. Besides exploring the potential for diagnostics, early works on two-photon-induced stimulated emission have been imperative for today's development of air-lasing concepts based on backward-directed stimulated emission. This chapter is devoted to a review and summary of these pioneering studies.

Since the invention of the laser in the early 1960s, a variety of laser-spectroscopic techniques have been developed for diagnostic purposes. These methods are extremely valuable for probing harsh environments such as combustion processes, where the nonintrusive nature and the high spatial and temporal resolution provided

J. Bood (✉) • M. Aldén
Department of Physics, Division of Combustion Physics, Lund University,
Box 118, 221 00 Lund, Sweden
e-mail: joakim.bood@forbrf.lth.se; marcus.alden@forbrf.lth.se

P. Polynkin, Y. Cheng (eds.), *Air Lasing*, Springer Series in Optical Sciences 208, https://doi.org/10.1007/978-3-319-65220-7_1

Table 1.1 Excitation and emission wavelengths for some species probed by two-photon LIF

Species	Excitation (nm)	Emission (nm)	Ref.
H	205	656	[4]
O	226	845	[5]
N	207/211	745/868	[6]
C	280	910	[7]
Ar	753/796	642	[8]
CO	230	400–600	[9]
NH_3	305	565	[10]
H_2O	248	400–500	[11]

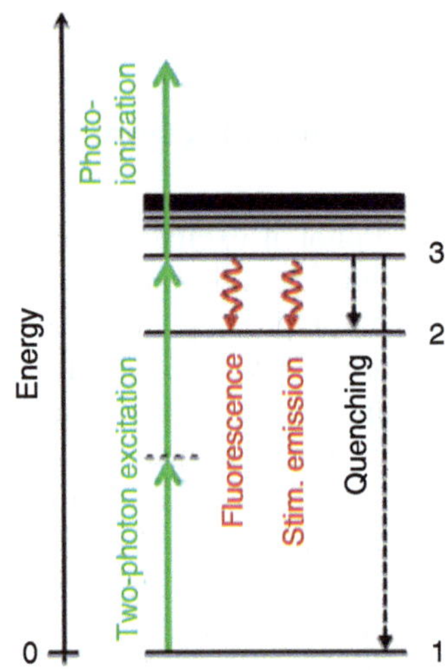

Fig. 1.1 Energy-level diagram and transitions involved in two-photon LIF

by optical techniques are particularly useful. Spontaneous Raman scattering and coherent anti-Stokes Raman scattering (CARS) are commonly used for the measurements of the concentrations and temperatures of the major species, while laser-induced fluorescence (LIF) has been widely used for the detection of trace species, such as extremely reactive and short-lived radicals. Comprehensive reviews of laser-spectroscopic techniques for combustion diagnostics are given in [1–3], and references therein.

Several species of great importance in the combustion and atmospheric research have absorption resonances in the vacuum ultraviolet (VUV) region, i.e., below 200 nm. Resonances in the VUV regime are inaccessible for probing in practical diagnostics under atmospheric conditions, since atmosphere strongly absorbs and scatters VUV radiation. Nevertheless, such species can be probed using excitation wavelengths above 200 nm via simultaneous absorption of two or more photons, i.e., multiphoton excitation. Table 1.1 lists important combustion species that have been detected with two-photon LIF.

A schematic illustration of two-photon LIF for detection of an atomic species is shown in Fig. 1.1. As can be seen in the figure, emission of fluorescence is not the only de-excitation channel. The atom can also be deactivated through collisional quenching, as well as stimulated emission (SE). In addition, the atom can be ionized through the absorption of an additional photon. In single-photon LIF, stimulated emission typically has a much lower rate than collisional quenching, and thus quenching is the major loss channel. In two-photon LIF, however, with the high laser intensities required to drive the excitation, SE can be a significant de-excitation channel, as a substantial population inversion between states 3 and 2 may be created. Therefore, several studies have been focused on how the two-photon LIF signal is impacted by SE [5, 12, 13].

It was soon realized that SE possesses some very attractive features for diagnostics, such as high-signal intensity and the fact that the signal is generated in a laser-like beam in both forward and background directions. SE potentially offers detection sensitivity exceeding that one of LIF and enables single-ended measurements, as only one-sided optical access is needed if the backward-generated SE is utilized. SE has not evolved into a major diagnostic tool due to various impairments inherent to the SE process, such as the integrative-growth nature of the signal and poor and sometimes ambiguous spatial resolution. However, in the recent years, the creation of population inversion and accompanying stimulated emission have become very interesting in the rapidly growing field of active remote sensing as means for producing backward-propagating lasing in air. The present chapter is devoted to the review of works on two-photon-induced SE for diagnostics. The focus is on oxygen and nitrogen atoms due to their relevance to air lasing, and CO, as this is perhaps the most thoroughly investigated specie in terms of diagnostic properties of SE. For the investigations focused on other species, the reader is referred to the following papers: [14, 15] for hydrogen, [16] for carbon, [17] for chlorine, [18] for krypton and xenon, and [10, 19] for ammonia.

1.2 Detection of Oxygen Atoms Through Two-Photon Excitation

A schematic energy-level diagram of the oxygen atom with relevant transitions indicated is shown in Fig. 1.2. Excitation from $2p^3$P to $3p^3$P, i.e., between states within the same Rydberg series, is done through the simultaneous absorption of two 226 nm photons. De-excitation to the intermediate $3s^3$S state is dipole-allowed and may occur via spontaneous emission, i.e., fluorescence at 845 nm. Collisions may induce a non-radiative transition to the $3p^5$P state, from which fluorescence at 777 nm may be emitted upon relaxation to the $3s^5$S state. Since single-photon transition from $3p^3$P to the ground state $2p^3$P is not dipole-allowed (the two states are of the same parity), an inverted population may be created between the pumped $3p^3$P state and the unpopulated $3s^3$S state. Population inversion can also be created between $3p^5$P and $3s^5$S states, given that $3p^5$P gets collisionally populated. SE can

Fig. 1.2 Energy-level diagram (not to scale) of atomic oxygen with the relevant transitions indicated

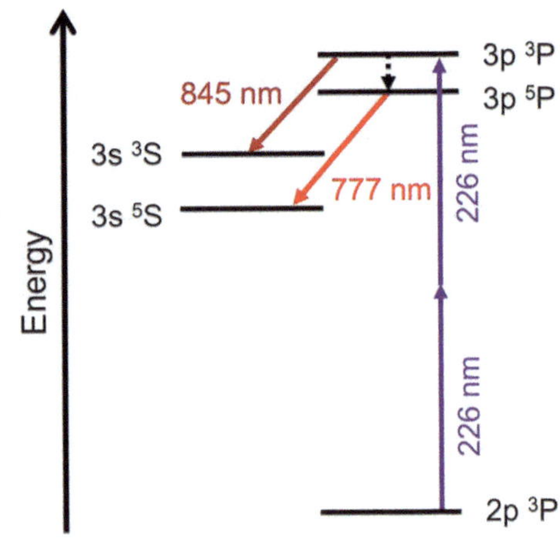

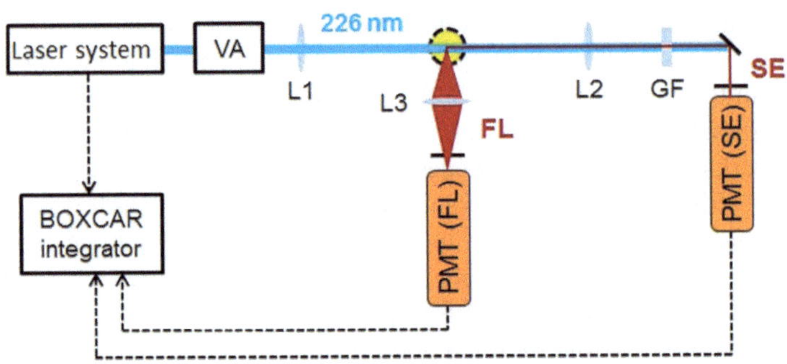

Fig. 1.3 Schematic setup for the detection of oxygen atoms through two-photon LIF and SE. The abbreviations are as follows: *VA* variable attenuator, *FL* fluorescence, *GF* glass filter, *SE* stimulated emission, and *PMT* photomultiplier tube

occur provided that the laser intensity is high enough to establish a population density in the excited state above threshold for single-pass SE.

A schematic experimental setup for the detection of oxygen atoms through two-photon excitation is depicted in Fig. 1.3. A frequency conversion of the output of a dye laser generated 5 ns pulses at 226 nm wavelength with a pulse energy of 3 mJ and a full width at half maximum (FWHM) linewidth of ~1 cm^{-1}. A variable attenuator (VA) allowed the laser pulse energy to be varied before the laser beam was focused into the sample volume by a 25 cm focal length spherical lens (L1). Fluorescence (FL) was collected perpendicularly to the direction of the laser beam and detected by a photomultiplier tube (PMT). Stimulated emission in the forward direction was collimated with a 25 cm focal length spherical lens (L2) and discriminated from the laser beam by a glass filter (GF) before being detected by a PMT. Three different samples were investigated: a low-pressure H_2/O_2 flame and free flows of room temperature N_2O and O_2. While naturally present oxygen atoms

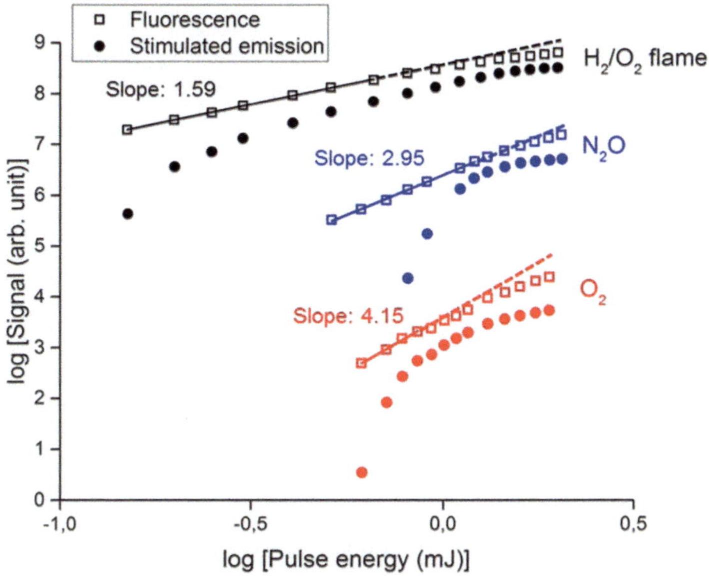

Fig. 1.4 Laser-energy dependence of fluorescence and SE from oxygen atoms recorded in three different environments (Adapted from Aldén et al. [20])

were detected in the flame, the oxygen atoms detected in the two gas flows were produced through photolysis by the 266 nm laser pulse.

It was observed that the forward-propagating signal, i.e., the SE, was more than four orders of magnitude stronger than the signal emitted to the side, i.e., the fluorescence. Furthermore, spectral investigation of these emissions revealed that the fluorescence spectrum exhibits two peaks, a stronger one at 845 nm, corresponding to the triplet-triplet transition, and a weaker one at 777 nm, corresponding to the quintet-to-quintet transition. The SE spectrum, however, only showed a single peak at 845 nm corresponding to the triplet-triplet transition. The absence of the quintet emission (777 nm) in the SE spectrum indicates that the laser pulse does not create a sufficient population inversion between the quintet states to reach threshold for SE.

In order to unravel different dependences of fluorescence and SE on the density of excited atoms, the emissions were studied as functions of the laser intensity, with three different samples. The results of those measurements are shown in Fig. 1.4. As can be seen, for all three cases, the low-intensity behavior of the SE is clearly different from that of the fluorescence, indicating that SE has a threshold, whereas the fluorescence data follow a linear curve in this regime. The solid lines represent the least square fits to the low-intensity fluorescence data, while the dashed lines are the extrapolations to the high-intensity region. The slopes of these log-log plots are indicative of the cumulative multiphoton orders involved in the pumping of population inversion. The slopes of the fits are 1.59 for the H_2/O_2 flame, 2.95 for the N_2O, and 4.15 for the O_2. The slope of 2.95 for the case of N_2O is consistent with

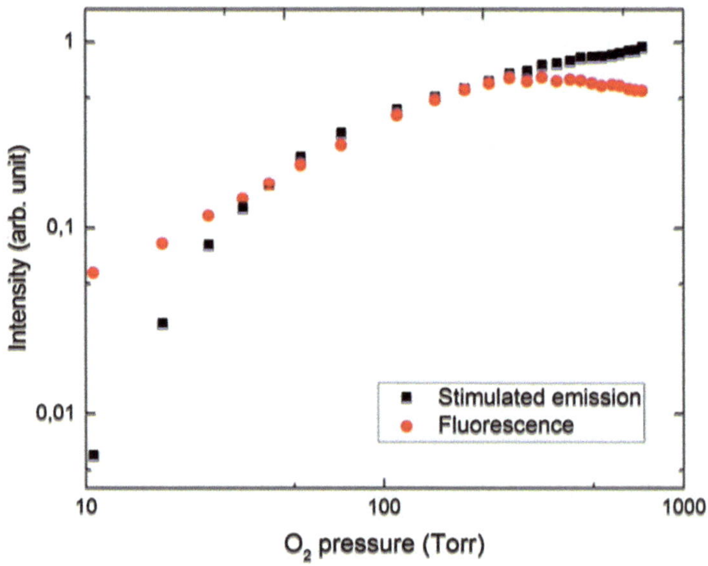

Fig. 1.5 Pressure dependences of the fluorescence and SE signals (Adapted from Aldén et al. [20])

single-photon photolysis followed by two-photon excitation of the oxygen fragment. For the case of O_2, the slope of 4.15 is consistent with two-photon photofragmentation of O_2 followed by two-photon excitation of the produced oxygen atom. The slope of 1.59 in the case of the H_2/O_2 flame is lower than the expected value of 2. Rapid single-photon photo-ionization and/or partial saturation of the two-photon transition may be the reasons for the reduction of the observed slope. These effects could also be responsible for the deviation from the low-intensity behavior at high pulse energies, indicated by the difference between the dashed lines and the data points shown in Fig. 1.4. It is also quite possible that rapid depletion of the two-photon-excited state by SE contributes to the discrepancy.

Data for pressure dependence of fluorescence and SE signals, recorded in pure O_2, are shown in Fig. 1.5. At low pressure the slope of the fluorescence signal in the log-log plot is close to unity, showing that it scales linearly with pressure. The SE signal on the other hand shows a much higher slope, indicating nonlinear pressure dependence. The fluorescence signal falls smoothly toward zero with decreasing pressure, while the SE signal drops abruptly at around 30 Torr, indicating threshold with respect to population inversion that is needed to generate SE.

Temporal dynamics of stimulated emission from oxygen atoms in a H_2/O_2 flame have been studied using 226 nm pump pulses of 10 ps duration and a streak camera. The results are shown in Fig. 1.6. It is evident from the data that the leading front of the signal pulse follows that of the pump pulse. The following part of the signal, however, decays due to the exponential dependence of the signal on the population inversion, which in turn has an exponential dependence on the quenching rate.

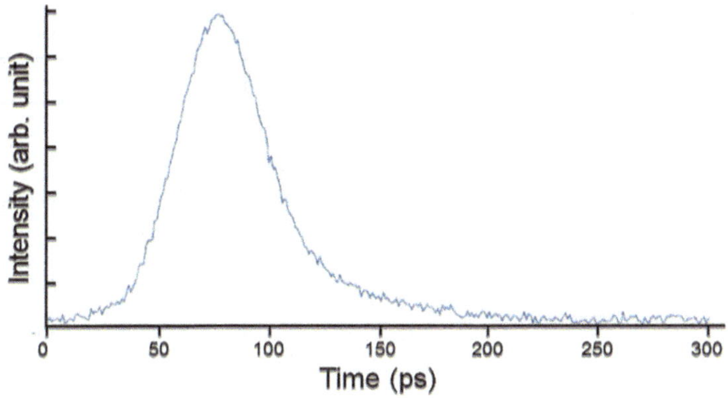

Fig. 1.6 Stimulated emission from oxygen atoms in a H_2/O_2 flame measured by a streak camera. 10 ps pulses of 226 nm wavelengths were used for the two-photon excitation (Reproduced from Agrup et al. [21])

1.3 Detection of Nitrogen Atoms Through Two-Photon Excitation

A schematic energy-level diagram of the nitrogen atom, with relevant transitions indicated, is shown in Fig. 1.7. Two-photon pumping from the $2p^4S^0$ ground state to the $3p^4D^0$ state, using an intense laser pulse of 211 nm wavelength, creates population inversion between this state and the $3s^4P$ state, resulting in SE at 870 nm.

The experimental setup for the studies of SE from nitrogen atoms was similar to the setup for oxygen atoms shown previously (see Fig. 1.3) and is described in detail in [22]. A dye laser, operated with a DCM dye solution and pumped by a 532 nm radiation from a Nd:YAG laser, was tuned to 633 nm. This radiation was frequency-doubled, and the resulting second-harmonic light was sum-frequency mixed with the 633 nm radiation to produce radiation at 211 nm. The 211 nm laser beam is isolated from the residual radiation at 633 and 316.5 nm using a Pellin-Broca prism. Continuous adjustment of pump power while maintaining the spatial profile and the direction of the beam was provided by a variable attenuator. The pump beam was focused into the probe volume using spherical lenses with focal lengths of 200 or 500 mm. The SE, generated in the wavelength range 868–872 nm, was re-collimated with a spherical lens and spectrally discriminated against the primary laser beam using an RG-695 Schott filter, before focusing onto the entrance slit of a spectrograph with another spherical lens. The spectrally dispersed SE was analyzed with an optical multichannel analyzer (OMA) system based on a time-gated diode-array detector. The signal was attenuated by two to three orders of magnitude using neutral-density filters in order to maintain a linear detection regime. The measurement volume was either an atmospheric premixed one-dimensional NH_3/O_2 flame on a water-cooled porous plug burner or NH_3 gas in a stainless steel cell equipped with quartz windows and connected to a vacuum and gas-flow systems.

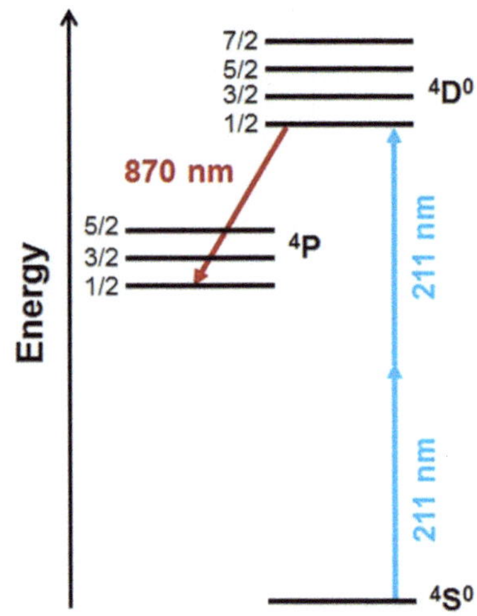

Fig. 1.7 Energy-level diagram (not to scale) of atomic nitrogen with the relevant transitions indicated

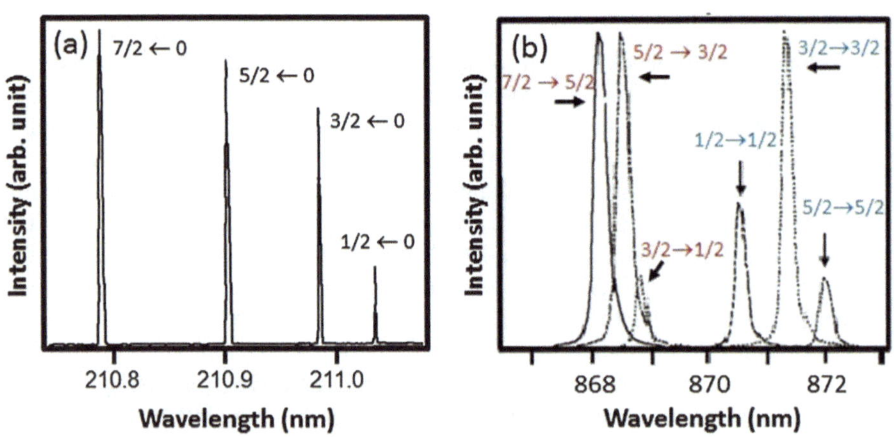

Fig. 1.8 Spectra of excitation (**a**) and stimulated emission, SE (**b**) from nitrogen atoms recorded in an NH_3/O_2 flame (Reproduced from Agrup et al. [22])

Figure 1.8a shows the SE excitation spectrum recorded in a fuel-rich (equivalence ratio $\phi = 1.6$) NH_3/O_2 flame, using 1.5 mJ of laser pulse energy focused by an f = 500 mm lens to a point located 1 mm above the burner. The four peaks correspond to the absorption transitions to the four fine-structure components of the upper $^4D^0$ state (see Fig. 1.7) with the $0 \rightarrow 7/2$ transition being strongest. The corresponding emission spectra, recorded in the same flame, are shown in Fig. 1.8b. The figure shows four different emission spectra, corresponding to the excitation using four different absorption lines shown in Fig. 1.8a. As seen in the emission spectrum (panel b), only transitions with $\Delta J = 0$ (blue designations) and $\Delta J = -1$

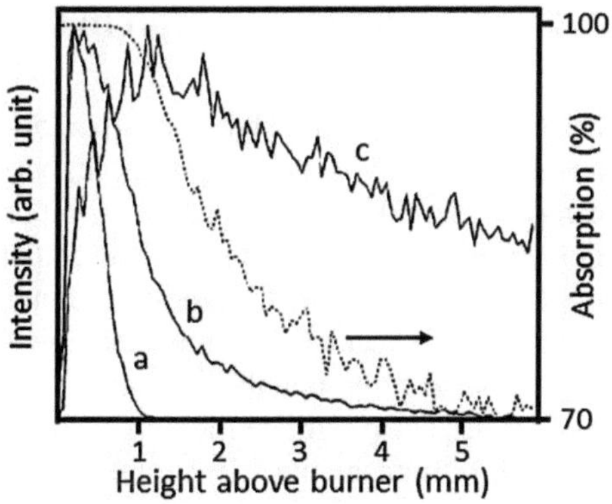

Fig. 1.9 Signal intensity vs. height above the burner for NH_3/O_2 flames with $\phi = 0.8$ (*a*), 1.0 (*b*), and 1.6 (*c*). The *dotted line* shows the UV laser power absorption characteristics for the richest flame ($\phi = 1.6$) (Reproduced from Agrup et al. [22])

(red designations) are observed. The reason why the lines corresponding to $\Delta J = +1$ are missing in the spectra is most likely due to their significantly lower transition probability [23]. Only the transitions that originate from the directly pumped energy level are observed in SE, while fluorescence spectra also reveal the emission peaks originating from other fine-structure levels populated through the collisional energy transfer [23].

To assess the potential of the SE signal for combustion diagnostics, the SE emission from nitrogen atoms was recorded at different heights above the burner and in different flame compositions. Fig. 1.9 shows the normalized signal profiles acquired in NH_3/O_2 flames with equivalence ratios $\phi = 0.8$ (a), 1.0 (b), and 1.6 (c).

The laser pulse energy was 1.9 mJ and the beam was focused with an f = 500 mm spherical lens. The dashed curve shows the absorption profile measured, in the flame with $\phi = 1.6$. It is noteworthy that this curve, which reflects the NH_3 concentration, does not correlate with the corresponding SE signal (c), which passes through the entire flame. This observation suggests that the SE signal does not originate from the photodissociation of NH_3. Another possible precursor to nitrogen atoms is nitric oxide (NO) that is present in relatively high concentrations (on the order of 10,000 ppm) in NH_3 flames [24]. However, measurements in a cell filled with NO, having approximately the same concentration as in the flame, i.e., 10^{16} molecules/cm^3, and Argon bath gas, have not resulted in any SE signal [22]. Even with higher laser irradiance, established by the use of an f = 200 mm focusing lens and 1.5 mJ laser pulse energy, no SE was observed.

Since the signal originates from a stimulated process, it can be further enhanced by an optical feedback. That was confirmed by reflecting a fraction of the SE back into the gain medium, i.e., the volume carrying population inversion, resulting in signal amplification. It was also observed that SE is emitted both in the forward and in the backward directions. The fact that a laser-like beam is emitted in the backward direction is of course particularly attractive as it opens up a pathway toward single-ended diagnostics, where only one-sided optical access is needed. That functionality is, above all, absolutely essential for the actively investigated sensing approaches based on air lasing.

1.4 Stimulated Emission in Carbon Monoxide

Molecular specie that exhibits two-photon-pumped fluorescence of the type discussed above for atomic oxygen and nitrogen, but without the need for the dissociation step is carbon monoxide CO. This gas is an important specie in the atmospheric and combustion chemistry. Its relevant energy-level diagram is shown in Fig. 1.10. Two photons of the wavelength 230 nm drive the transition from the $X^1\Sigma^+$ ground state to the $B^1\Sigma^+$ excited state. Radiation is then be emitted through the relaxation to different vibrational levels in the $A^1\Pi$ state or through the $b^3\Sigma$ - $a^3\Sigma$ system upon collisional energy transfer (Q).

In order to investigate the diagnostic potential of SE for the detection of CO and how SE influences the LIF signal from CO, a setup in which forward-propagated and backward-propagated SE could be fed back into the probe volume was constructed. The setup is shown in Fig. 1.11. Laser pulses (13 ns FWHM) at 230 nm

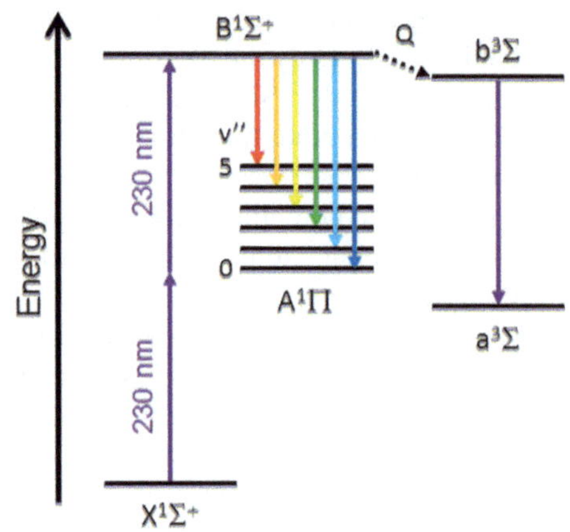

Fig. 1.10 Energy-level diagram for two-photon LIF and SE of CO

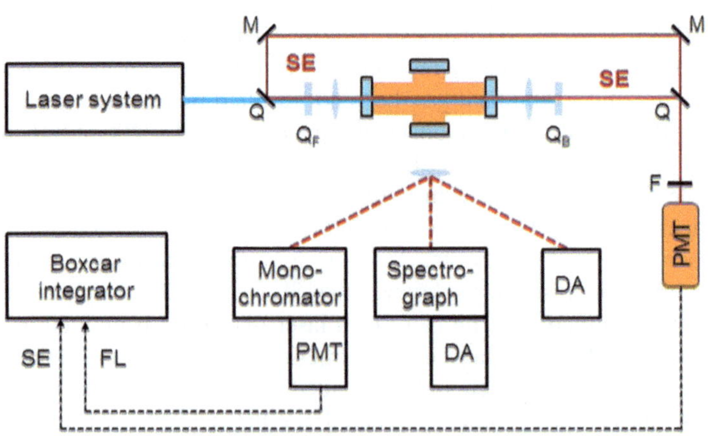

Fig. 1.11 Experimental setup for two-photon LIF and SE studies in carbon monoxide

wavelength were generated using an Nd:YAG/dye laser system, where the frequency-doubled output of the dye laser was mixed with the residual radiation at 1064 nm to generate laser radiation at 230 nm, with 3.5 mJ of energy per pulse. The CO gas was contained in a stainless steel pressure cell equipped with entrance and exit windows made of fused silica as well as a separate perpendicularly oriented quartz window for fluorescence studies. Fused silica slides were placed before and after the cell to add feedback in the forward and backward directions. The slide labeled Q_F added feedback to the forward-propagating SE, while the slide marked Q_B added feedback to the backward-propagating SE. The SE, attenuated with neutral-density filters (F), was monitored with a PMT, while the fluorescence emitted sideways could be monitored in the spatially resolved fashion, using a diode-array detector (DA), or in the spectrally resolved fashion, using a monochromator or a spectrograph. The fluorescence and SE signals registered by the two PMTs were directed to a boxcar integrator for simultaneous processing. Additional experimental details are given in [13].

While the fluorescence signals associated with both the B-A and b-a systems were observed, SE was only observed for the transitions belonging to the B-A system. Furthermore, only three transitions, corresponding to the largest Franck-Condon factors, were observed in the SE spectrum, as shown in Fig. 1.12. In the linear fluorescence signal, also the peaks associated with the transitions with lower Franck-Condon factors were observed [13]. Figure 1.12 also shows the SE spectra recorded for the cases with feedbacks. The spectra are normalized to unity at the highest peak. As can be seen, the relative intensities of the peaks are changed when feedback is added. The feedback enhances weaker ro-vibrational transitions relative

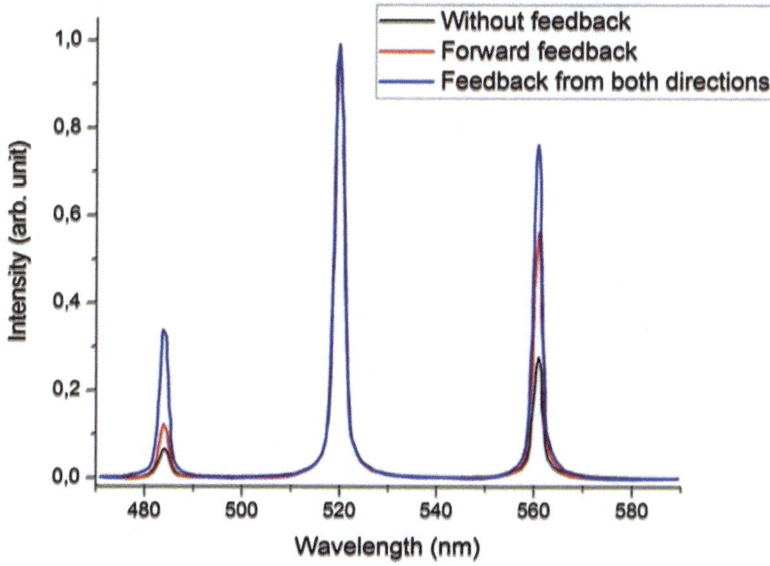

Fig. 1.12 Spectra of the SE signals recorded without feedback, with forward feedback only, and with feedback from both directions (Reproduced from Westblom et al. [13])

to the stronger ones. The reason for this is that stronger transitions saturate sooner than weaker transitions as the feedback is introduced, and the SE signal becomes stronger. With the laser intensity close to threshold for SE, which was 0.4 mJ with a focusing lens of 200 mm focal length and cell pressure of 80 Torr, feedback enhanced the SE signal by more than two orders of magnitude.

Experiments were carried out to study the relative strengths of the SE signals emitted in the forward and backward directions. Stimulated emission signals were recorded for various CO pressures, operating with 1–2 mJ/pulse and using two different lenses to focus the laser beam (focal lengths of 200 mm and 500 mm). When the longer focal length lens was used, the SE intensities in the forward and backward directions were found to be approximately equal. With the shorter focal length lens, it was found that the SE signal generated in the backward direction was about a factor of two more intense than that generated in the forward direction. This forward-backward asymmetry seemed to be independent of pressure in the range 5–100 Torr, and rather independent of pulse energy, once the process was well above threshold. The difference in the behavior for the two focal length lenses may be due to greater contributions of other nonlinear optical processes in the higher-intensity focal region produced by the shorter focal length lens.

Figure 1.13 shows pressure dependence of the fluorescence and SE signals with and without feedback. As can be clearly seen in the figure, the two signals behave very differently as pressure is increased. The SE signal increases rapidly followed

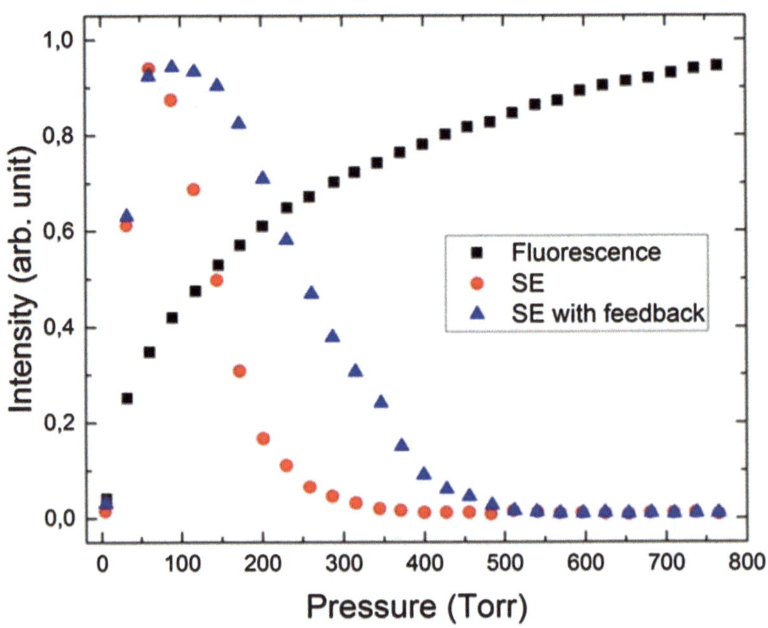

Fig. 1.13 Fluorescence and SE signals from CO versus pressure (Reproduced from Westblom et al. [13])

by an almost equally rapid decrease, while the LIF signal increases smoothly. Furthermore, the addition of feedback shifts the falling slope of the SE curve toward higher pressure. This behavior is reasonable since the initial value of the SE rate is higher with feedback than without, allowing the stimulated process to build up.

Although one can envisage combustion diagnostic schemes that rely on the detection of SE, a more traditional diagnostic approach relies on the detection of LIF, which has an advantage of being a position-specific quantity. In this regard, the generation of SE can be a detrimental effect. As can be seen in Fig.1.1, SE is a depopulation process that competes with the signal of interest, i.e., the fluorescence. While the third deactivation channel, i.e., the collisional quenching, has been the subject of numerous studies (see, e.g., [25] and references therein), far fewer studies of the influence of SE have been undertaken. Since it is not practically possible to turn the SE process on and off, the impact of SE was investigated by studying the changes in the fluorescence intensity and spectral content while changing the strength of the SE, which could be made by introducing optical feedback. Figure 1.14 shows a CO fluorescence spectrum recorded with and without feedback. The signals are spectrally virtually identical, but the addition of feedback reduces the fluorescence intensity by approximately 25%. The reduction of the LIF signal increases with laser intensity, but is independent of the CO pressure. Hence, these spectra show that the introduction of feedback affects the LIF signal. However, they do not tell anything about the absolute scale of the influence by SE on the fluorescence signal in the absence of feedback.

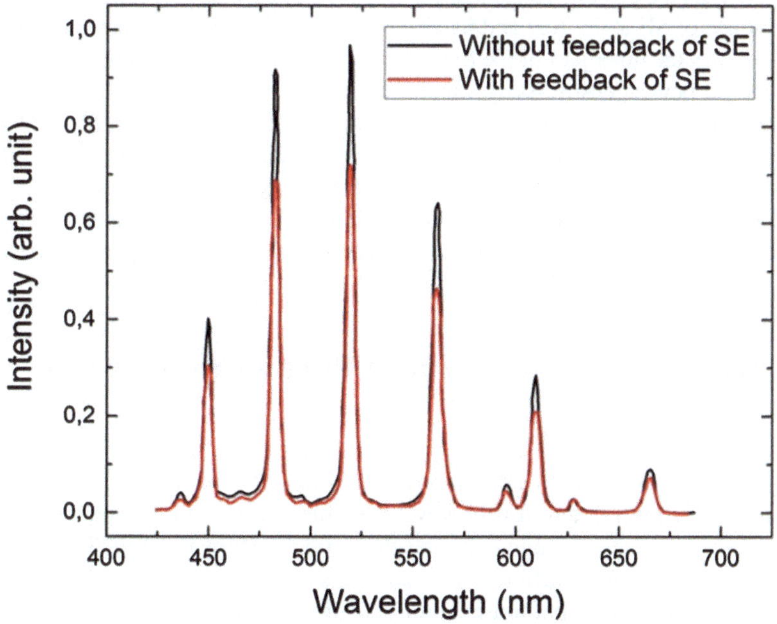

Fig. 1.14 Laser-induced fluorescence spectrum of CO with and without feedback of SE (Reproduced from Westblom et al. [13])

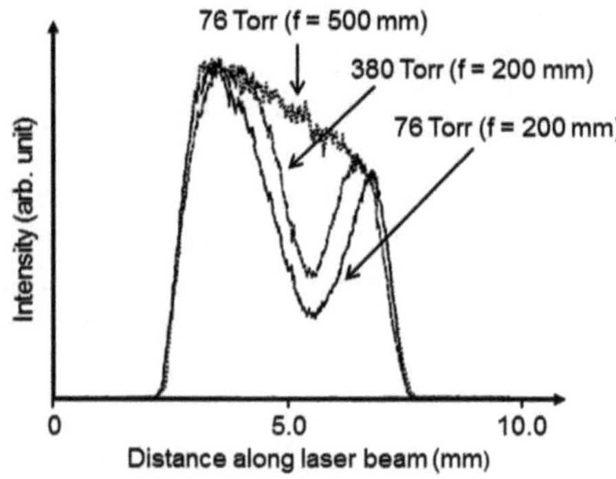

Fig. 1.15 Normalized spatially resolved fluorescence signals from CO recorded at 76 Torr (strong SE) and 380 Torr (virtually zero SE). The fact that the dip in the profile is deeper at 76 Torr than at 380 Torr suggests that SE influences the fluorescence signal (From Westblom et al. [13])

In order to investigate the effect of SE on the fluorescence signal without adding feedback, a spatially resolved study of the fluorescence was conducted at two different pressures, 76 Torr and 380 Torr. According to Fig. 1.13, SE is strong at 76 Torr, while it is essentially absent at 380 Torr. The results of the fluorescence measurements are displayed in Fig. 1.15. Ideally, the spatially resolved fluorescence signal recorded with a weakly focused laser beam (f = 500 mm) should be constant along the sample, but due to the absorption of the laser light, the curve for the fluorescence signal vs. coordinate is tilted. The dip observed in the spectra recorded with tighter focusing of the laser beam (f = 200 mm) is mainly due to photo-ionization.

As the data in Fig. 1.15 clearly show, the profile recorded at 76 Torr has a deeper minimum than the profile recorded at 380 Torr. We believe that this difference is due to the presence of SE at 76 Torr and its absence at 380 Torr. Furthermore, the difference occurs only in the inner part of the focus zone, due to the threshold behavior of SE with respect to the power density. As a result, SE occurs in a narrower region of the interaction region compared to fluorescence. Frank et al. have studied the radial profiles of the two-photon LIF signals from oxygen atoms excited with picosecond laser pulses of different intensities in an H_2/O_2 flame [5]. They found that the spatial distribution of the signal is essentially constant for low excitation intensities, but it becomes significantly modified at highest laser intensities, due to the combined influence of SE and photo-ionization.

1.5 Improved Spatial Resolution Using a Crossed-Beam Setup

A major drawback of SE as a diagnostic tool is its poor spatial resolution along the direction of the laser beam. It is however possible to achieve high spatial resolution using a crossed-beam arrangement as schematically shown in Fig. 1.16 [26]. The two beams have slightly different frequencies such that the sum-frequency exactly matches that of the two-photon excitation of interest. The beams are spatially displaced and focused so that they cross inside the gas sample. As a result, population inversion is only established in the overlap region, and a strong SE signal is emitted in the direction of the bisector of the two crossing beams.

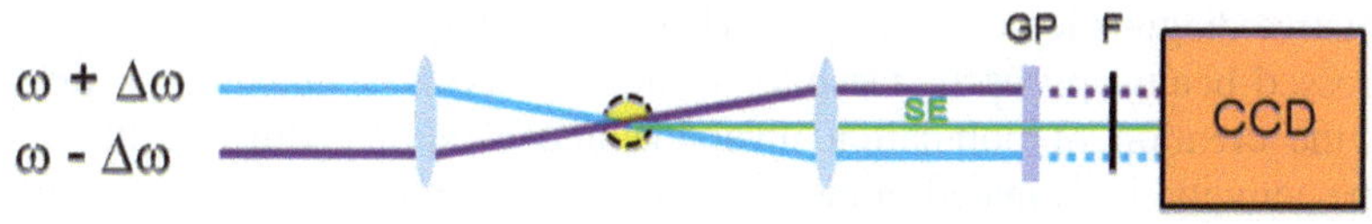

Fig. 1.16 Crossed-beam arrangement for spatially resolved diagnostics with SE. SE designates the stimulated emission signal, GP is a glass plate in which the primary laser beam radiation is absorbed, while the SE signal is transmitted, and F indicates neutral-density filters

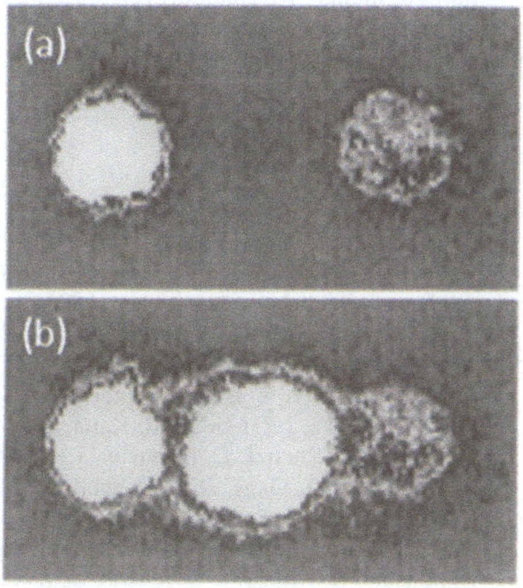

Fig. 1.17 Images of SE recorded with a CCD camera without (**a**) and with (**b**) NH_3 present in the probe volume. As can be clearly seen in (**b**), a strong beam of SE is present between the two primary laser beams (Reproduced from Giorgiev et al. [27])

This method has been applied to the detection of NH_3. Figure 1.17a shows an image recorded with a CCD camera without NH_3 present in the probe volume, while Fig. 1.17b shows the corresponding image recorded with NH_3 present. As clearly seen in panel (b), a strong beam of SE is generated between the two primary laser beams.

Brown and Jeffries have demonstrated a similar approach in which quantitative oxygen concentrations have been determined by monitoring gain for an additional probe beam, whose wavelength was tuned to the transition corresponding to the SE, crossing the two-photon excitation beam at an angle [27].

1.6 Concluding Remarks

Although stimulated emission possesses several attractive features, it has not evolved into a major diagnostic tool. This is mainly due to several difficulties inherent of the SE process, namely, the integrative-growth nature of the signal and poor and sometimes ambiguous spatial resolution. These characteristics render quantitative diagnostics very difficult. These difficulties can be overcome by using the crossed-beam approach. Despite its limitations as a quantitative diagnostic itself, the creation of population inversion and accompanying SE has recently become actively investigated in the context of active remote sensing as a mean for producing backward lasing in air. Quoting Dogariu and Miles [28]: "The air laser opens the way for remote optical detection of molecular species by means of directional coherent backwards emission coming from the target."

Acknowledgments We gratefully acknowledge the contributions of the present and former members of the Division of Combustion Physics at Lund University and our collaboration partners, particularly Ulf Westblom, Sara Agrup, and John Goldsmith.

References

1. C. Eckbreth, *Laser Diagnostics for Combustion Temperature and Species*, 2nd edn. (Gordon and Breach Publishers, Amsterdam, 1996)
2. K. Kohse-Höinghaus, J.B. Jeffries, *Applied Combustion Diagnostics* (Taylor and Francis, New York, 2002)
3. M. Aldén, J. Bood, Z. Li, M. Richter, Proc. Combust. Inst. **33**, 69 (2011)
4. W.D. Kulatilaka, J.R. Gord, V.R. Katta, S. Roy, Opt. Lett. **37**, 3051 (2012)
5. J.H. Frank, X. Chen, B.D. Patterson, T.B. Settersten, Appl. Opt. **43**, 2588 (2004)
6. E. Wagenaars, T. Gans, D. O'Connell, K. Niemi, Plasma Sources Sci. Technol. **21**, 042002 (2012)
7. U. Westblom, P.-E. Bengtsson, M. Aldén, Appl. Phys. B Lasers Opt. **52**, 371 (1991)
8. H. Matsuta, K. Kitagawa, Spectrosc. Lett. **45**, 13 (2012)
9. J. Brackmann, J. Sjöholm, M. Rosell, J. Richter, M. Bood, Aldén, Proc. Combust. Inst. **34**, 3541 (2013)
10. C. Brackmann, B. Hole, Z.S. Zhou, M. Li, Aldén, Appl. Phys. B **115**, 25 (2014)
11. K. Larsson, O. Johansson, M. Aldén, J. Bood, Appl. Spectrosc. **68**, 1333 (2014)
12. Y.-L. Huang, R.J. Gordon, J. Chem. Phys. **97**, 6363 (1992)
13. U. Westblom, S. Agrup, M. Aldén, H.M. Hertz, J.E.M. Goldsmith, Appl. Phys. B Lasers Opt. **50**, 487 (1990)
14. J.E.M. Goldsmith, J. Opt. Soc. Am. B **6**, 1979 (1989)
15. J. Amorim, G. Baravian, M. Touzeau, J. Jolly, J. Appl. Phys. **76**, 1487 (1994)
16. M. Aldén, P.-E. Bengtsson, U. Westblom, Opt. Comm. **71**, 263 (1989)
17. A.D. Sappey, J.B. Jeffries, Appl. Phys. Lett. **55**, 1182 (1989)
18. J.C. Miller, Phys. Rev. A **40**, 6969 (1989)
19. U. Westblom, M. Aldén, Appl. Spectrosc. **44**, 881 (1990)
20. M. Aldén, U. Westblom, J.E.M. Goldsmith, Opt. Lett. **14**, 305 (1989)
21. S. Agrup, M. Aldén, Opt. Comm. **113**, 315 (1994)
22. S. Agrup, U. Westblom, M. Aldén, Chem. Phys. Lett. **170**, 406 (1990)
23. J.B. Jeffries, R.A. Copeland, D.R. Crosley, J. Chem. Phys. **91**, 2200 (1989)
24. V. Brackmann, A. Alekseev, B. Zhou, E. Nordström, P.-E. Bengtsson, Z. Li, M. Aldén, A.A. Konnov, Combust. Flame **163**, 370 (2016)
25. K. Kohse-Höinghaus, J.B. Jeffries, *Applied Combustion Diagnostics*, Chapter 5 (Taylor and Francis, New York, 2002), pp. 128–154

26. N. Georgiev, K. Nyholm, R. Fritzon, M. Aldén, Opt. Comm. **108**, 71 (1994)
27. M.S. Brown, J.B. Jeffries, Appl. Opt. **34**, 1127 (1995)
28. A. Dogariu, R. Miles, Backwards nitrogen double lasing in air for remote trace detection, in *Laser Applications to Chemical, Security and Environmental Analysis (LACSEA) 2014 Technical Digest*, (The Optical Society of America, Washington, DC, 2014.) paper: LW2D.3

Chapter 2
High-Gain Air Lasing by Multiphoton Pumping of Atomic Species

Arthur Dogariu and Richard Miles

2.1 Introduction

The possibility of achieving high-gain lasing from a remote location in air opens up numerous opportunities for the development of new concepts. Backward lasing has generated the most interest, since it may provide a high-sensitivity method for the detection of greenhouse gases, gas leakage from pipelines and refineries, pollution, illicit chemical and nuclear processing activities, chemical gas attacks, and the presence of explosives and hazardous materials [1, 2]. Other applications of high-gain air lasing are of significant interest and include "around-the-corner" illumination, clandestine communication, and a local "guide star" for the correction of aero-optical distortion.

The approach we have taken is to utilize the very high gain available from multiphoton pumping of atomic species. To date the atomic species have been created by the dissociation of major constituents of air, including oxygen [3–6], nitrogen [7–11], and water vapor [12, 13]. In each of these cases, the molecular species are dissociated, and the lasing is achieved by two-photon pumping of the atomic fragments. In some cases, the dissociation and two-photon pumping steps can be achieved by the same laser pulse; however, higher conversion efficiency is attained with a leading pulse providing dissociation followed by the two-photon pumping.

It is anticipated that high-gain lasing may also be possible using inert atomic gas species naturally present in air including argon, krypton, neon, and even xenon. In these cases, no dissociation step is necessary, so pulse-to-pulse fluctuations associated with the variations in the breakdown process are eliminated, and the stand-off detection through the use of modulation methods becomes more tractable. Argon is of particular interest since it is present in air at concentrations approaching 1%.

A. Dogariu (✉) • R. Miles
Department of Mechanical and Aerospace Engineering, Princeton University, Princeton, NJ 08544, USA
e-mail: adogariu@princeton.edu; miles@princeton.edu

P. Polynkin, Y. Cheng (eds.), *Air Lasing*, Springer Series in Optical Sciences 208, https://doi.org/10.1007/978-3-319-65220-7_2

However, it requires three-photon pumping due to the transparency limitations of air. Thus, even though the argon concentration in air is similar to the normal water vapor concentration in humid air, which has been successfully demonstrated to lase, lasing from argon will require significantly higher intensity pumping and/or longer-gain path length. This may be achievable with femtosecond pulses, which have been recently demonstrated to pump backward lasing from argon in a mixture of 10% argon in air [14, 15].

In all these cases, lasing occurs from a population inversion that is created in the focal volume of the pump laser, and the lasing direction is determined by the geometry of that volume, reflecting the exponential amplification of stimulated emission with path length. Due to the multiphoton nature of the pumping, the gain volume is well localized to the high-intensity region of the pump. In the normal case with standard lenses, that gain volume is elongated in the propagation direction of the pump laser and localized to the Rayleigh range of the laser focal volume. This leads to lasing in the forward and backward directions, with the backward-propagating lasing beam retracing the pump path backwards and the forward-propagating lasing overlapping with the continued propagation of the pump beam in the forward direction. Through the use of cylindrical focusing, the focal volume geometry can be changed such that the maximum gain path is in directions other than forward and backward, leading to "around-the-corner" lasing possibilities. Further localization and amplification of the air laser can be achieved with secondary gain regions that are timed and positioned to overlap the propagating air laser pulse [16].

These laser configurations are "mirrorless" or "cavityless" [17], similar to other high-gain lasers such as soft x-ray lasers [18], free electron lasers, and high-gain excimer and nitrogen lasers. For typical lasers, the spatial and temporal coherence of the laser beam owes much to the resonator cavity. The reflecting mirrors localize stimulated emission to a single- or low-order transverse mode of the resonator cavity, and the resonant standing wave modes of the cavity determine the linewidth and temporal structure of the output laser pulse(s). In the case of the cavityless laser, the transverse spatial mode of the laser beam echoes the pump volume and cross section, and temporally the emission is a single pulse or burst of pulses, each of which corresponds to a single gain pass through the medium. If the pump laser persists for times longer than a hundred picoseconds or so, then multiple lasing pulses can be created through the repeated depletion and re-pumping of the gain region. For both the normal laser and the cavityless laser, the spatial and temporal coherences can be well characterized. For cavityless lasers with high-gain media, low-order highly coherent modes dominate the amplified spontaneous emission background. In this case, there is no measurable difference between the pulse generated from a cavityless laser and a single pulse selected from the pulse train exiting from a mode-locked laser or a regenerative laser cavity. To verify that similarity, experiments were conducted to determine the spatial and temporal coherence and pulse characteristics of the air lasers.

2.2 Properties of Atomic Air Lasing

Figure 2.1 shows the atomic energy levels associated with atomic hydrogen, nitrogen, and oxygen lasers. Hydrogen is two-photon pumped with 205 nm light, nitrogen with 207 or 211 nm light, and oxygen with 226 nm light. Lasing can be achieved with picosecond and nanosecond pumping and, as will be discussed later, with femtosecond pumping. The general configuration for the experiments is shown in Fig. 2.2, indicating that both backward- and forward-propagating lasing emissions are monitored, as well as the emission to the side. Figure 2.3 shows the approximate geometry of the gain region, which is formed by the focal zone of the pump laser. A large L/d ratio ensures that the stimulated emission grows exponentially in the forward and backward directions, in the detriment of the side emission. Furthermore,

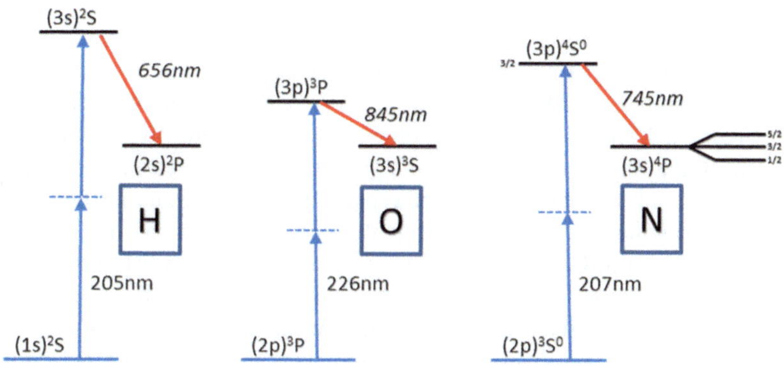

Fig. 2.1 Energy levels for two-photon pumping of hydrogen, oxygen, and nitrogen atoms

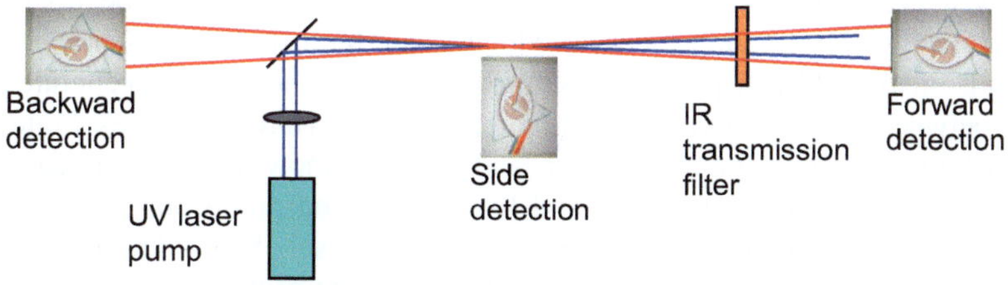

Fig. 2.2 Experimental setup for the detection of forward- and backward-propagating air lasing

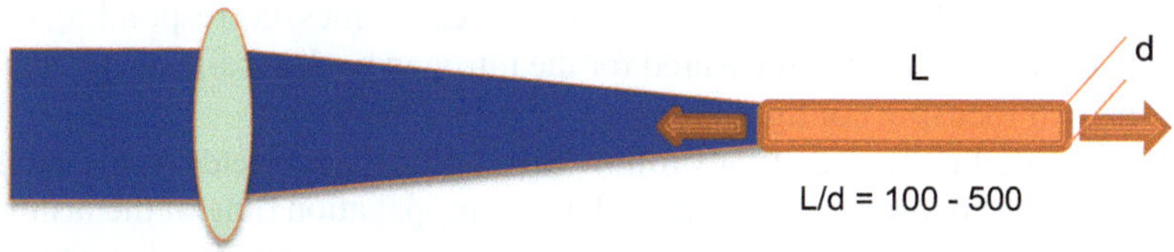

Fig. 2.3 Mechanism for air lasing: The geometry of the gain region allows for high gain along the propagation direction of the pump beam

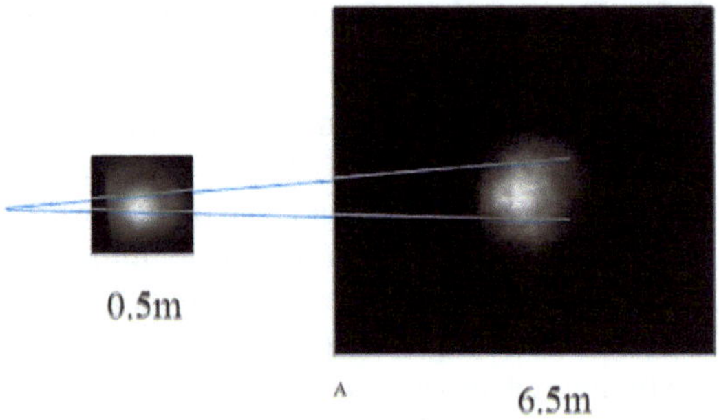

Fig. 2.4 Divergence of the air laser beam measurement for backward-propagating atomic nitrogen laser

the fast stimulated emission process depletes the excited states before much of the spontaneous emission can take place. Since the pump in all cases is in the ultraviolet and the lasing is in the red or infrared, the separation on the pump from the lasing is easily accomplished with dichroic mirrors or prisms.

The spatial coherence is determined by imaging the air laser beam profile, following the evolution of that profile with the distance from the air laser generating volume, and comparing that evolution of beam profile with that of an ideal Gaussian beam. Figure 2.4 shows the beam profile of the backward-propagating nitrogen laser measured at 0.5 and 6.5 m from the gain region. The measured divergence of 2 mrad corresponds to the divergence of a spatially coherent Gaussian beam. Similar results were obtained while monitoring the propagation of atomic oxygen and hydrogen laser beams. Comparing the divergence of the backwards air laser in oxygen with the focusing angle of the UV pump beam yields $\theta_{oxygen}/\theta_{UV} = 3.3$, very close to the 3.7 ratio between the 845 and 226 nm wavelengths for the air laser and pump beams, respectively. The divergence ratio indicates that the lasing region closely overlaps the pump volume.

The temporal coherence of atomic air lasing can be measured using a Michelson-Morley interferometer. The incoming laser beam is spilt into two beams that are recombined at the imaging plane, producing interference fringes. The coherence length is measured by displacing one mirror relative to the other, to generate a time delay between the two pulses while monitoring the interference fringes. Figure 2.5a shows the interferometer configuration, and Fig. 2.5b shows the interference fringes measured for the cases of tightly and loosely focused pump pulses. The tightly focused pump has a gain region of ~3 mm, and the loosely focused pump has a gain region of ~1 cm. Converting the propagation distances of the two interfering pulse to the time delay and plotting the fringe visibility as a function of this time delay provides a measure of the temporal coherence for these two configurations. Figure 2.5b shows 10 ps and 35 ps FWHM coherence times, corresponding to 3 mm and 1 cm coherence lengths measured for the nitrogen backward-propagating lasing under the two focusing configurations.

The measured coherence times follow the confocal parameter of the pump and indicate that the coherence is determined by the propagation time of the lasing pulse passing through the gain region. This coherence time of the air laser is matched very closely by the pulse length of the emitted pulses, indicating that the pulse duration

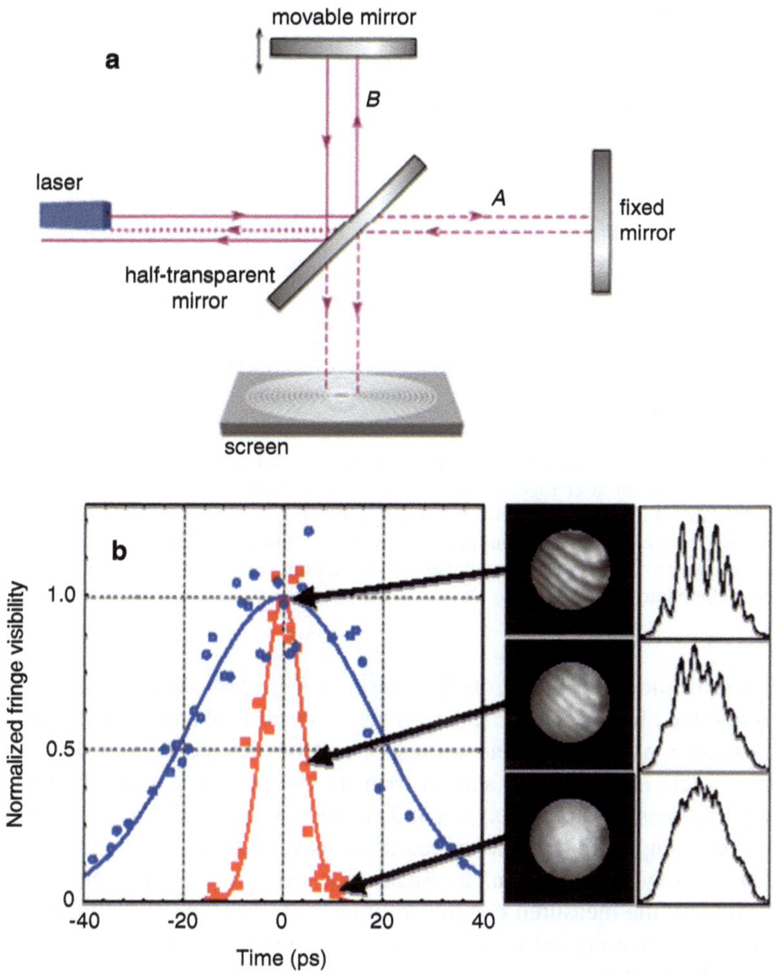

Fig. 2.5 Air laser temporal coherence measurements using a Michelson-Morley interferometer (**a**) interferometer configuration. (**b**) measured coherence time for 3 mm (*red*) and 1 cm (*blue*) focal volume lengths

of the air laser is close to the transform limit. Figure 2.6 shows results of direct measurements of the temporal profile of the air laser pulses for both oxygen and nitrogen lasers in the atmospheric air using a fast photodetector (20 ps response time) and a high-bandwidth oscilloscope (Tektronix DPO73304D 33 GHz Digital Oscilloscope).

Assuming Gaussian temporal pulse shape, the pulse length can be determined by the deconvolution of the detector response from the measured pulse waveform. Figure 2.6a shows the measurement of a sub-30 ps pulse for the oxygen laser emission. The measured pulse waveform of the nitrogen air laser, together with the ~20 ps response time of the measuring system, obtained using a 100 fs laser pulse

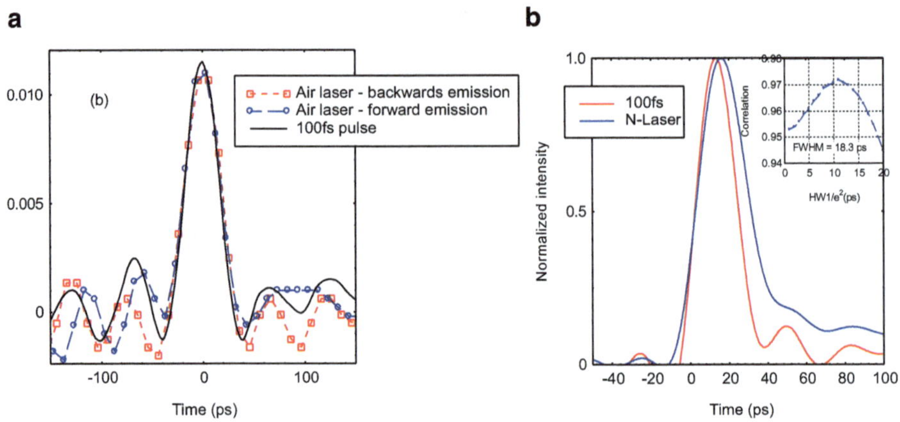

Fig. 2.6 (**a**) Sub-30ps-resolution measurement of the forward- and backward-propagating emissions of atomic oxygen laser together with the detector response measured using a 100 fs laser pulse; (**b**) temporal trace of the backward-propagating emission from the atomic nitrogen laser, together with the temporal resolution measured with a 100 fs pulse. The inset shows a 18.3 ps – long deconvolution of the measured pulse for the atomic nitrogen lasing from the temporal response of the photodetector

at 800 nm, is shown in Fig. 2.6b. The inset in Fig. 2.6b shows a pulse width of 18.3 ps obtained for the atomic nitrogen laser, which is obtained by deconvoluting the measured signal from the detector response. Note that the measured pulse widths are three orders of magnitude shorter than the 34 and 43 ns natural spontaneous emission lifetimes for atomic oxygen and nitrogen, respectively, and they scale with the gain path length. For all the air lasers we have studied, either from molecular dissociation (oxygen, nitrogen, hydrogen) or using preexisting atomic species (argon, xenon), the measured coherence times match very well with the measured pulse widths, indicating the transform limited nature of the backward and forward emitted air-lasing pulses.

The spectra shown in Fig. 2.7 correspond to lasing from the transitions depicted in Fig. 2.1. Hydrogen and oxygen atoms emit single spectral peaks 656 nm and at 845, respectively. The two emission lines in atomic nitrogen, at 744.23 and 746.83 nm, correspond to the transitions from $(3p)^4S^0_{3/2}$ state to the states $(3\ s)^4P_{1/2}$ and $(3\ s)^4P_{3/2}$, respectively. The spectra shown in Fig. 2.7 appear broader than the spectra of the actual emissions because of the limited resolution of the spectrometers used to record them.

2.3 Generation Mechanisms

Coherent light pulses can be generated by stimulated emission, super-fluorescence, or super-radiance. In all three cases, the emitted pulse can be spatially and temporally coherent, and the pulse length can be significantly shorter than the incoherent

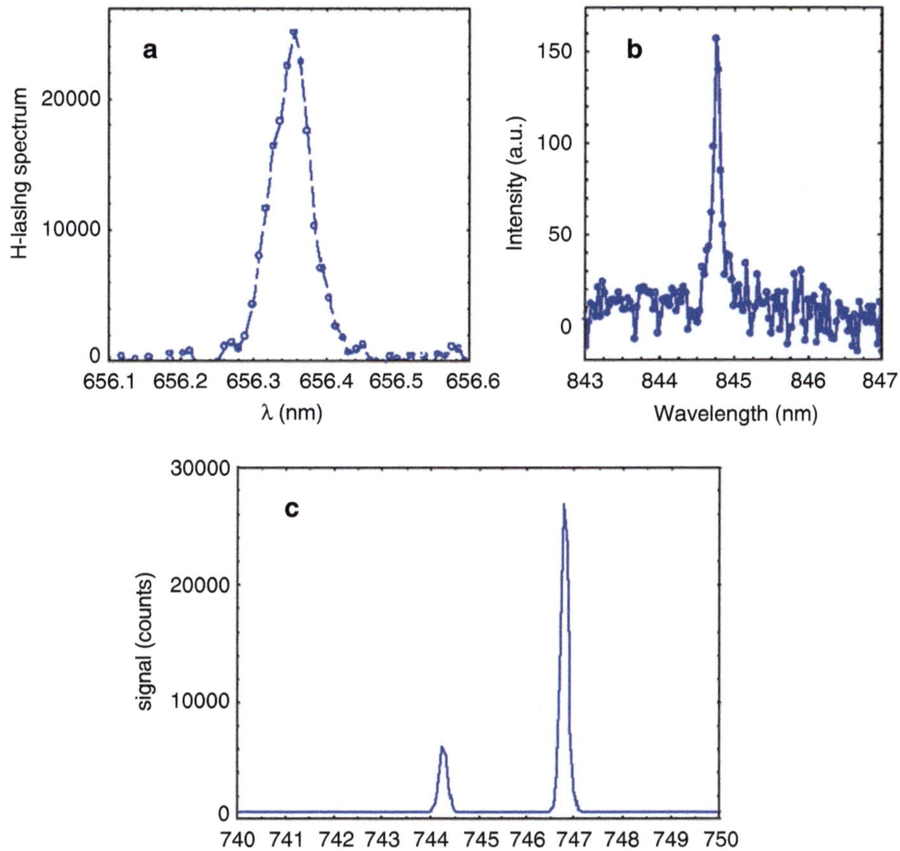

Fig. 2.7 Spectra of lasing in dissociated (atomic) hydrogen (**a**), oxygen (**b**), and nitrogen (**c**)

fluorescence emission lifetime. The difference is that the stimulated emission process has exponential gain and is unaffected by dephasing collisions, whereas super-radiance and super-fluorescence are associated with coherent emissions from multiple dipoles oscillating in phase. Thus, super-radiance and super-fluorescence scale quadratically with the number of dipoles involved in the emission and require the dipoles to be synchronously phased. The emission lifetimes of super-fluorescence and super-radiance are shorter than the fluorescence lifetime of a single dipole and are related to the number of coherently coupled dipoles radiating together [19]. The more dipoles are coupled, the faster the radiation process. Several experiments were conducted to establish the generation mechanism in atomic air lasers, including the measurements of the backward-propagating pulse energy as a function of the gain path length, pump energy, atom density, and dephasing time.

The gain profile was measured by monitoring the backward-propagating pulse energy as a function of the gain path length, by moving a beam blocking glass slide through the gain region. Figure 2.8 shows the results of that experiment for the

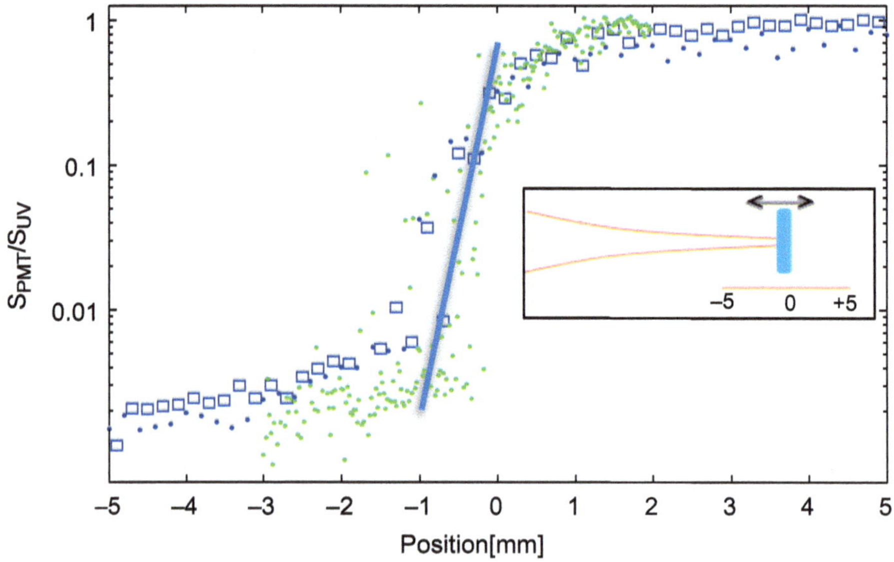

Fig. 2.8 Backward emission for atomic oxygen laser with varying length of the gain region. The slope of the fit indicates gain coefficient between 40–80 cm^{-1}

backward-propagating oxygen laser. The beam blocking glass slide is translated a total of 10 mm, and the gain region is measured to be approximately 1 mm, corresponding to the confocal beam parameter of the pump laser (twice the Rayleigh range). A fit to the gain profile over this 1 mm region assuming constant pump intensity gives an exponential gain of between 40 and 80 cm^{-1}. The line in the diagram has a slope corresponding to a gain of 60 cm^{-1}, an increase by a factor of e^6 over 1 mm.

Another measurement of gain can be extracted from the ratio of the emission from the side of the lasing volume (i.e., fluorescence) to the backward lasing emission. In that case, the side detector shown in Fig. 2.2 is used. Figure 2.9 shows the relative signals from oxygen measured in the backward and side directions. To determine the total fluorescence emission, the detected fluorescence signal was multiplied by the ratio of the solid angle covered by the detector area to 4π steradians. Figure 2.10 is a scatter plot of that measurement, with each point representing a simultaneously recorded separate pump and the associated fluorescence and lasing events. The scatter arises from the pump laser fluctuations.

The above measurement yields backward coherent emission 500 times stronger than the volume-integrated incoherent emission. Assuming exponential gain, e^{gL}, where L = 1 mm, the measurement yields a gain coefficient of 62 cm^{-1}, in good agreement with the measurement shown in Fig. 2.8. The very high optical gain and the high directionality (low divergence) lead to *six orders* of magnitude enhancement in energy per unit solid angle for backward-propagating coherent emission relative to the incoherent emission. Both this and the moving glass slide

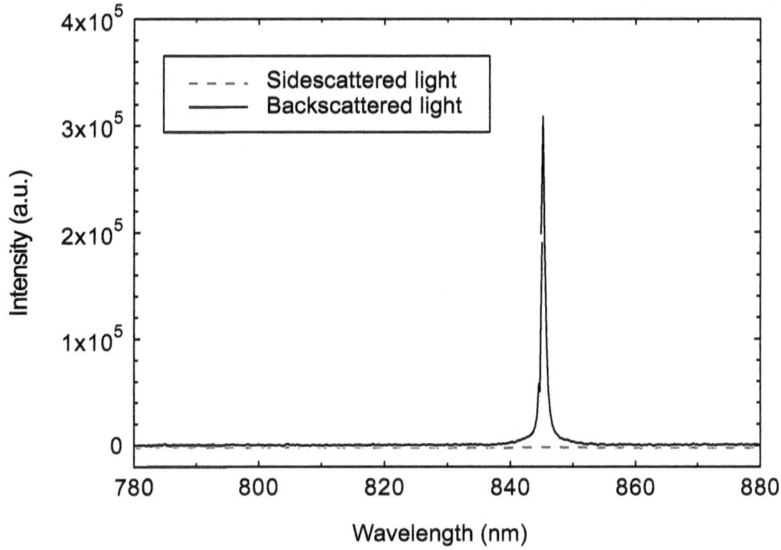

Fig. 2.9 Measured spectra of the backward laser emission ("backward detection" in Fig. 2.2) and of the spontaneous emission ("side detection" in Fig. 2.2), for atomic oxygen emission in air

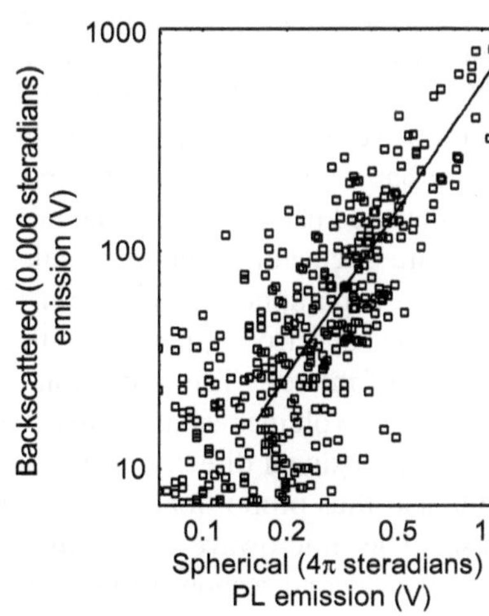

Fig. 2.10 Backward-propagating coherent lasing emission plotted as a function of the incoherent side emission integrated over the solid angle of 4π steradians

measurement indicate an exponential process producing the coherent forward- and backward-propagating beams.

Another indication of stimulated emission is the variation of pulse energy of the backward-propagating coherent beam with the pump energy. Figure 2.11 shows a scatter plot of the backward lasing pulse energy vs. pump pulse energy. Note that there is a clear threshold point where gain for the backward signal overcomes losses and backward lasing begins to dominate fluorescence. From that point, the backward lasing increases in a highly nonlinear fashion.

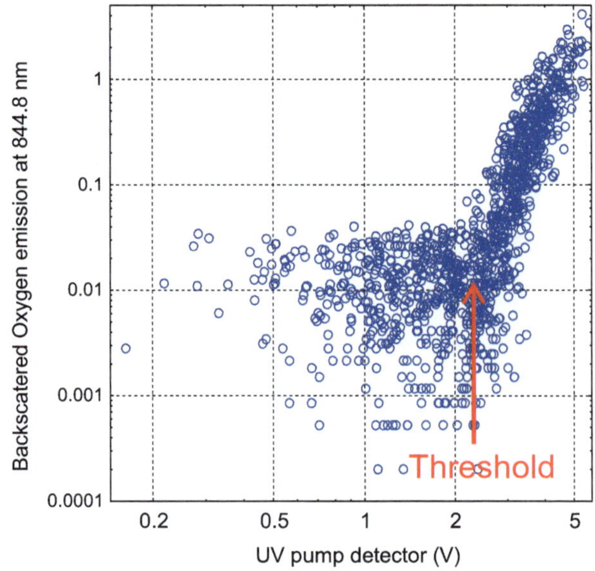

Fig. 2.11 The backwards emission from the atomic oxygen air laser as a function of the pump power indicates clear threshold behavior

To better quantify the nonlinear increase of the lasing signal with the pump pulse energy, a direct measurement of the number of pumped oxygen atoms, taken simultaneously with backward lasing, was undertaken using the Radar REMPI technique [20]. Figure 2.12 shows the experimental arrangement. A continuous low-power microwave beam illuminates the lasing region, and the microwave backscattering from that region is collected, separated from the outgoing microwave, amplified, and detected using a homodyne mixing. Radar REMPI has been previously used to measure trace atomic and molecule species in air. It takes advantage of resonance-enhanced multiphoton ionization [21] (hence the abbreviation REMPI), a well-developed spectroscopic method that relies on the selective ionization of particular species of interest. The ionization is monitored by coherent microwave scattering, which, for volumes smaller that the microwave wavelength (3 mm in this case), produces a signal directly proportional to the number of free electrons in the volume and thus proportional to the number of ions created by the multiphoton process. The microwave scattering occurs at the time of ionization and has sub-nanosecond response time, so it is not strongly affected by quenching and is immune to the background light from the laser scattering or fluorescence. Previous measurements with trace nitric oxide in air have shown excellent linearity of the Radar REMPI signal with the nitric oxide mole fraction [22]. For oxygen and other atomic species that lead to air lasing, the same two-photon resonance that pumps the lasing transition can be used as the resonant intermediate state for the REMPI diagnostic. This corresponds to 2 + 1 REMPI – two photons to couple the resonance and one more for ionization. Assuming only a small fraction of excited atoms are ionized, the Radar REMPI signal is directly proportional to the number of oxygen atoms reaching the excited two-photon resonant state. The REMPI provides a

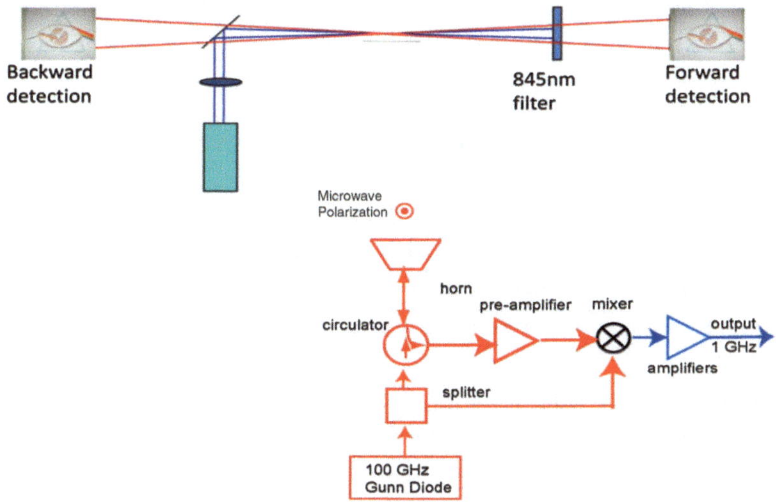

Fig. 2.12 The Radar REMPI setup (*bottom*) monitors the excited-state population by measuring the amount of ionization induced by the same multiphoton excitation responsible for air lasing (*top*)

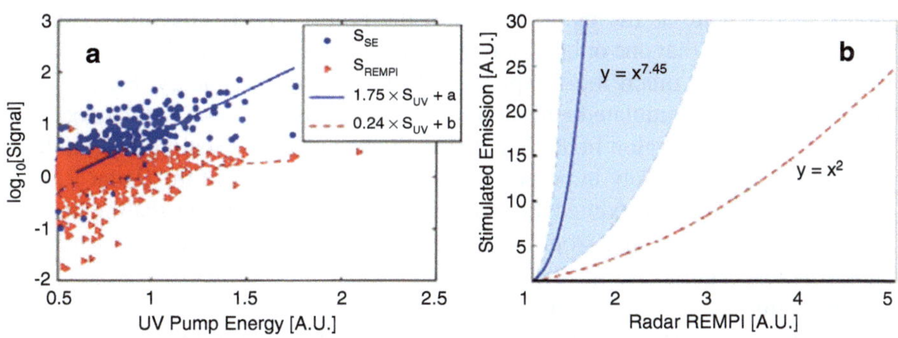

Fig. 2.13 (**a**) The backward emission signal (S_{SE}) shows stronger dependence on the pump power than the dependence of the excited-state density monitored by the Radar REMPI (S_{REMPI}). (**b**) Plot of the backward lasing as a function of the excited-state density shows a dependence closer to the seventh-order exponential (*solid line*, indicating stimulated emission), as opposed to the quadratic dependence (*dashed line*), which would be expected for super-radiance

means for measuring the scaling of the backward lasing pulse energy with respect to the population of excited atoms.

Figure 2.13a shows a scatter plot of the Radar REMPI signal and the backward lasing pulse energy as a function of the pump laser pulse energy. Data for the Radar REMPI and for the backward lasing are shown with red and blue colors, respectively. Note that the Radar REMPI signal is sublinear in the high pump pulse energy regime, indicating that the number of atoms excited in this regime is limited by

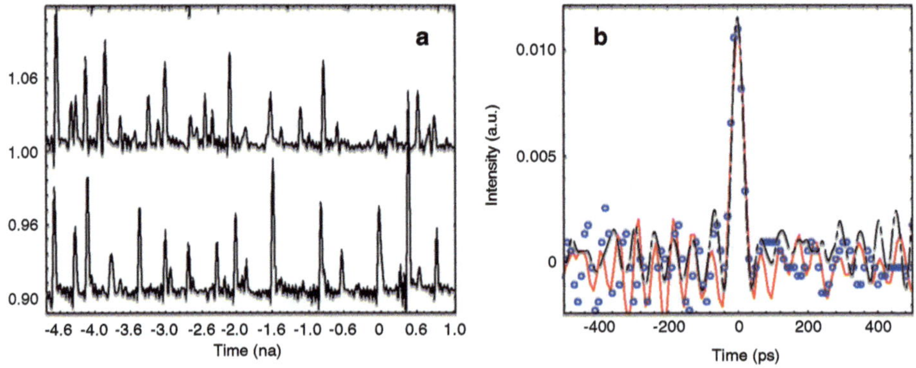

Fig. 2.14 (**a**) Oxygen air lasing under nanosecond pumping shows correlated forward (*upper curve*) and backward (*lower curve*) emissions. (**b**) A zoom-in showing the temporal pulse shapes of the individual forward- and backward-propagating laser pulses within the emission burst

something other than the laser intensity. In this experiment, the same laser pulse was used to dissociate the oxygen molecules, and only following the dissociation step, there are oxygen atoms available for the two-photon pumping. Both the dissociation and the pumping processes require two pump photons, so the number of excited atoms should scale as the pump energy to the fourth power. The slowly varying increase indicates that one or both of these processes are saturated. Stimulated emission clearly scales much more rapidly with the laser pulse energy. Figure 2.13b shows the ratio of stimulated emission to the Radar REMPI, with the data occupying the blue shaded region in the figure. The fit shows that the scaling of stimulated emission is approximately in proportion to the seventh power of the Radar REMPI and, by assumption, the excited atom density. This is a much faster scaling than the quadratic relationship that would be expected in the cases of super-radiance and super-fluorescence.

Air lasing can be easily obtained with long pump pulses, albeit using higher pulse energy in order to maintain high enough intensity for multiphoton excitation. Using nanosecond pulses for pumping oxygen or nitrogen backward lasing provides sufficient molecular dissociation within the pumping pulse, such that a pre-dissociating pulse is not required. Since stimulated emission depletes the excited state in a matter of 10–20 ps, continuous pumping for several nanoseconds provides the opportunity to generate multiple pulses via the succession of multiphoton excitation and rapid stimulated emission. Figure 2.14a shows bursts of picosecond pulses obtained through air lasing in oxygen in forward and backward directions under nanosecond optical pumping. The 10 ns UV pump pulse continuously excites the atoms via multiphoton excitation, and successive picosecond pulses are observed in both forward and backward directions. These pulses are detected simultaneously with fast detectors and constitute highly correlated trains of separate 10–20 picosecond pulses. The number of these pulses and the time between them, which is required to build up the gain again after each lasing pulse, are both strong indicators of the mechanism leading to the observed coherent emission. Examples of individual forward- and backward-propagating pulses are shown in Fig. 2.14b, where both the forward and backward pulses have identical temporal durations of less than 20 ps.

An intrinsic feature of stimulated emission is that the gain is related to the population inversion and is only weakly affected by dephasing collisions. On the contrary, super-radiance and super-fluorescence are directly affected by the dephasing collisions since both of those coherent gain mechanisms rely on the radiating dipoles being prepared in phase for the coherent radiation to be emitted. Thus, another approach to determine the physics underlying the emission is to vary the dephasing collision frequency by changing the pressure of the gas.

In order to test collision effects, a simple experiment with a 10 ns pump pulse was performed to study air lasing while varying the de-coherence time. The pressure of the air was varied over more than an order of magnitude, allowing the dephasing time (through collisions) to vary from more than 100 ps to less than 10 ps. As seen in Fig. 2.15, the lasing is robust with respect to pressure variations, and the emission is retained even when the dephasing time is shorter than the pulse width of each lasing spike.

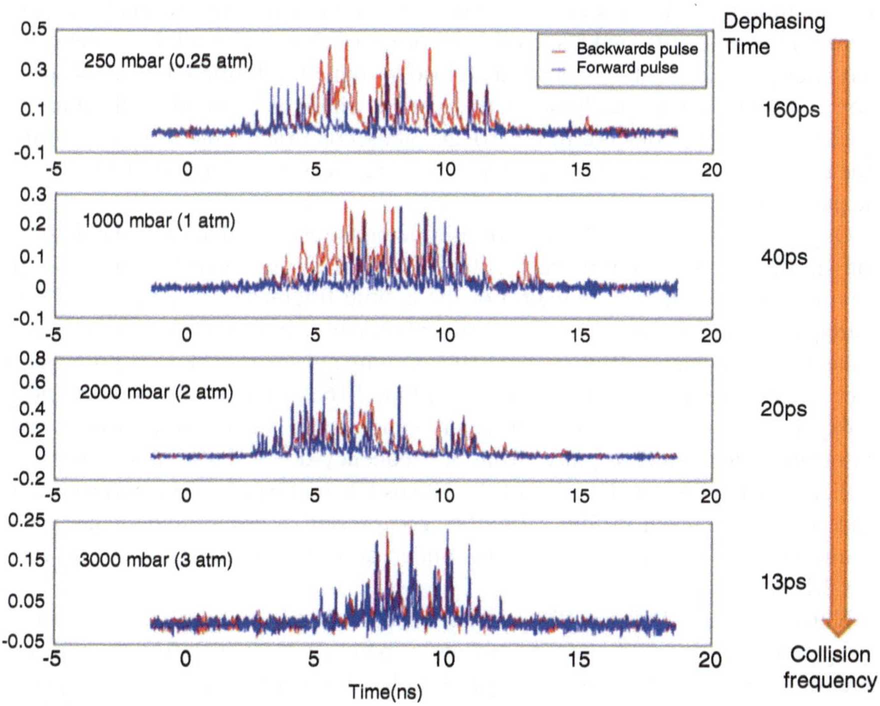

Fig. 2.15 Varying the collision frequency (through changing air pressure) leaves the temporal pulse bursts emitted by the nanosecond-pumped air laser qualitatively unchanged

Taken together, the exponential gain, the much faster than quadratic scaling of the emission with the atom density, the insensitivity to dephasing, and the scaling of the pulse durations with the duration of the single passage through the gain region unambiguously indicate that the backward lasing process is associated with stimulated emission rather than super-fluorescence or super-radiance. Simultaneous emission in the forward and backward directions from gain volumes with lengths corresponding to thousands of optical wavelengths is another indication of the dominance of the stimulated emission process. We point out that our conclusion that stimulated emission is responsible for the generation of the lasing in air is at odds with the conclusion reached by some other authors [23, 24] and the experiments presented here were conducted to settle that argument.

2.4 Lasing Through Multiphoton Pumping of Molecular Species

Lasing processes in atomic oxygen, nitrogen, and hydrogen, discussed above, have many similarities as well as some important differences. All three gain media require dissociation of the corresponding molecular species followed by resonant two-photon pumping to the upper emission state. All three require ultraviolet laser pump at specific wavelengths and lase in the red or near infrared. For all three, the generated laser pulse is highly coherent and close to transform limited. Lasing occurs as soon as the gain overcomes losses, which can be within picoseconds of the pumping process. Pulse durations are in the 10–20 picosecond range and reflect the transit time through the gain region – longer pulses are measured with looser focusing of the pump laser.

Oxygen was the first high-gain air laser demonstrated. In that case, the dissociation step could be accomplished with the same 100 picosecond laser pulse that provided the two-photon excitation of the atomic fragment, since there is a good overlap of the 226 nm energy with the two-photon transition to the dissociative state of the oxygen molecule. With a single 100 ps pump pulse, the backward-propagating laser mode had a donut shape, as shown in Fig. 2.16a. A factor of 50 higher energy of oxygen lasing, as well as a Gaussian lasing mode, was observed when a pre-dissociating nanosecond pulse from a frequency-doubled Nd:YAG laser was applied, few microseconds ahead of the 100 ps UV pump pulse. The corresponding mode profile is shown in Fig. 2.16c. For comparison, a lasing mode generated in atomic oxygen in methane/air flame under atmospheric pressure is shown in Fig. 2.16b.

Nitrogen pumped with a 100 picosecond UV laser required a pre-pulse for the lasing to occur, due to a stronger molecular bond (9.78 eV bonding energy compared to 5.16 eV in the case of oxygen molecule) and the lack of a two-photon overlap with a dissociative state accessed by either 205 or 211 nm pump light. In Fig. 2.17a, b, we show how a pre-pulse, few nanoseconds earlier than the pulse that

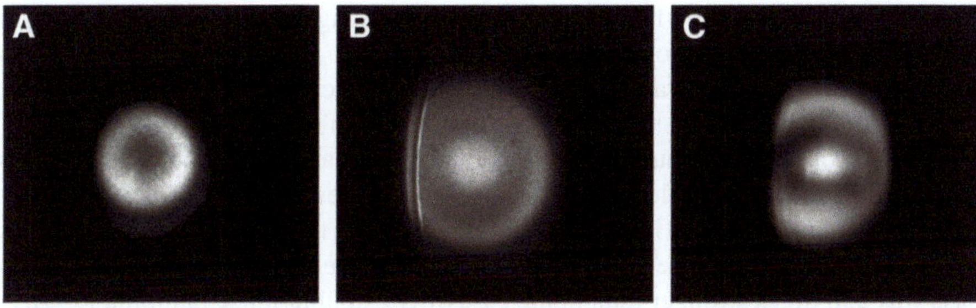

Fig. 2.16 Spatial mode profiles emitted by the atomic oxygen laser in the backward direction, in the cases when (**a**) molecular oxygen in the air is dissociated with the same single 100 ps pulse that pumps the atomic fragments, (**b**) atomic oxygen is dissociated in a methane/air flame, and (**c**) atomic oxygen is produced through optical breakdown of air

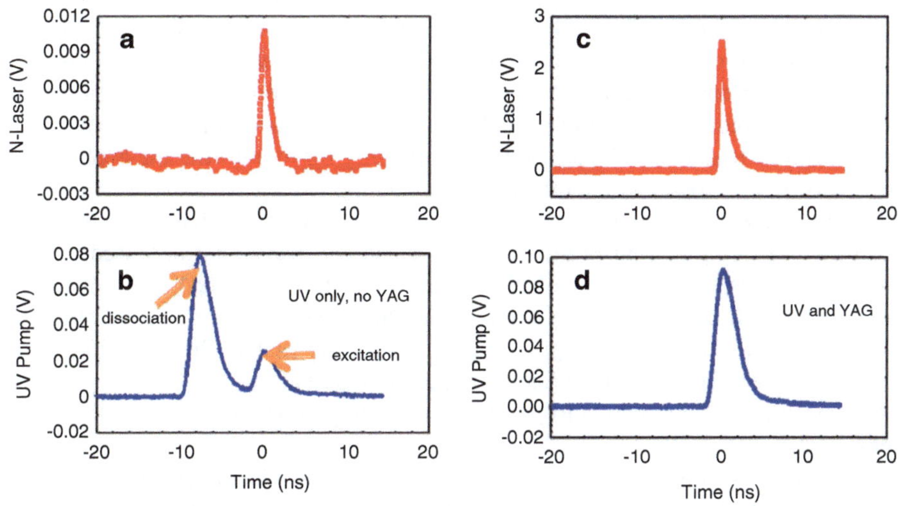

Fig. 2.17 (**a**) Nitrogen lasing pumped with a pair of UV pulses. (**b**) The corresponding temporal waveform of the dual pump pulse. (**c**) A two-order magnitude of stronger nitrogen lasing pumped by a pre-dissociating Nd:YAG laser pulse, followed by a UV pump pulse. (**d**) The corresponding YAG + UV pump waveform

pumps the atomic fragments, can lead to lasing from atomic nitrogen in air. The first pulse, which is produced by the same UV pump laser, dissociates the nitrogen molecule through a multiphoton process, while the second pulse drives the two-photon excitation of the atomic fragments. No lasing is observed from the first pulse alone even though it is at the two-photon excitation wavelength. When both pulses in the sequence are present, the appearance of the lasing emission temporally coincides with the timing of the second pump pulse, indicating that the separate dissociation step was necessary, in contrast to the case of oxygen.

Much stronger output from atomic nitrogen lasing can be achieved using a higher-energy pre-pulse, optimally timed before the two-photon UV pump. In Fig. 2.17c, d, we demonstrate that the lasing threshold can be reduced significantly, and, hence, the lasing emission increased by more than two orders of magnitude, if a strong nanosecond pre-pulse from a Nd:YAG laser (at either 1064 or 532 nm) is used to dissociate the molecules in advance of the resonant UV pulse. The pre-pulse, which is not shown in Fig. 2.17d, was sent 100 ns in advance of the 100 ps UV pulse, to dissociate the nitrogen molecule. Nitrogen emission obtained in this case was 250 times stronger than in the case of UV double-pulse pumping [7–9]. A similar excitation approach with a nanosecond pre-pulse has been investigated for both oxygen and nitrogen and nanosecond DUV pumping [10] and for oxygen with femtosecond DUV pumping [25].

By optimizing the delay between the pre-pulse and the DUV pumping pulse, almost three orders of magnitude enhancement was obtained in the case of atomic nitrogen. The lasing is the strongest when the highest density of atomic nitrogen is present and available to be excited by the two-photon resonant UV pump pulse. In order to monitor the atomic density, we utilize the fact that the same photons used for the two-photon nitrogen atomic lasing excitation also produce two-photon resonant three-photon ionization of a small fraction of the nitrogen atoms. In this case, time-delayed Radar REMPI provides the capability for monitoring the evolution of the density of nitrogen atoms following dissociation.

In Fig. 2.18, we show the dependence of the atomic nitrogen density on the delay between the femtosecond pre-dissociating pulse and the signal amplitude of the resonant UV-pumped Radar REMPI. As shown with red squares in Fig. 2.18, the nitrogen atomic density reaches a maximum within the first microsecond after dissociation in the atmospheric pressure air, indicating that is the best timing for achieving the most efficient air lasing. In pure nitrogen at the atmospheric pressure, the concentration of atomic nitrogen resulting from photodissociation is greater and peaks later.

The time evolution of the atomic density measured via Radar REMPI can be directly compared with the dynamics of air lasing, since both processes utilize the same two-photon resonance. Figure 2.19 shows that the temporal evolution of the gain coefficient (given by the natural logarithm of the backward lasing signal) closely follows the atomic density measured with Radar REMPI and plotted on a linear scale. The fact that the exponential gain coefficient is proportional to the atomic density is another strong indication of stimulated emission.

Both oxygen and nitrogen lasing arise from the two major species in the air, while hydrogen lasing occurs from the water vapor in the air at as low as 40% relative humidity, corresponding to the mole fraction below 1%. Figure 2.20a shows the relative backward hydrogen lasing pulse energy as a function of humidity in the room air at 25 °C temperature. When pumped by a 100 ps UV laser pulse, a preliminary dissociation step is required. In these experiments, the water molecule was dissociated using either a 10 nanosecond Nd:YAG laser producing 200 mJ pulses at 1064 nm or a 50 fs Ti:Sapphire laser generating 1 mJ pulses at 800 nm.

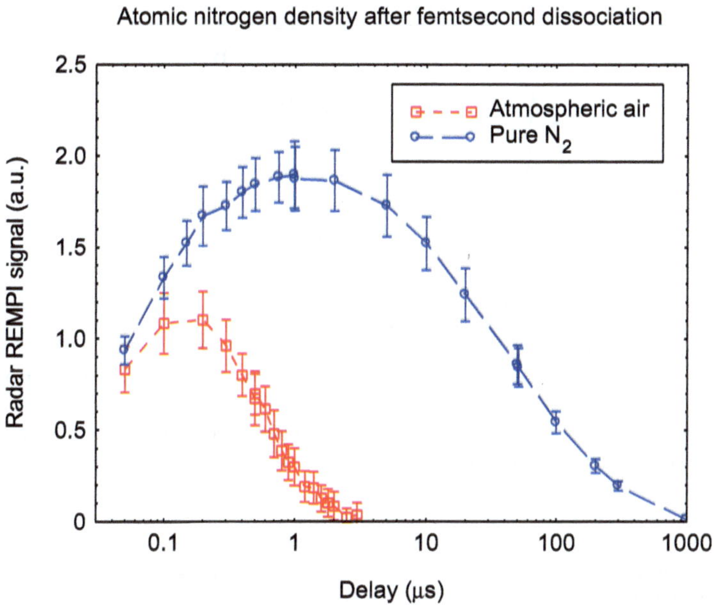

Fig. 2.18 Density of atomic nitrogen as a function of the delay between molecular dissociation and atomic excitation

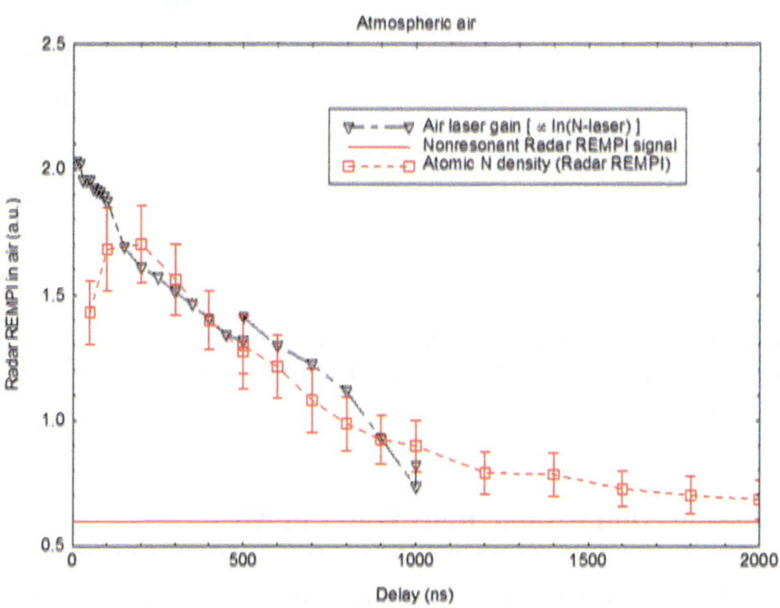

Fig. 2.19 Temporal evolution of the gain coefficient, measured as the natural logarithm of the backward emission and shown with inverted triangles, closely follows the dynamics of the excited state of atomic nitrogen, measured by Radar REMPI and shown with squares

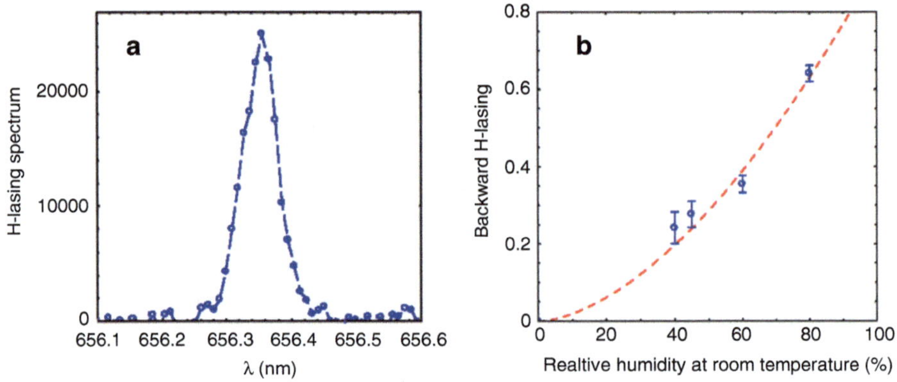

Fig. 2.20 (**a**) Spectrum of backward-propagating hydrogen lasing. (**b**) Pulse energy of the hydrogen lasing in air as a function of humidity showing a nonlinear dependence of the emission on the concentration of water molecules

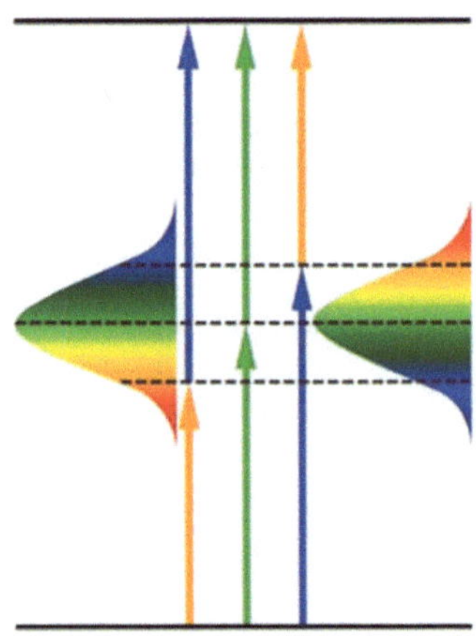

Fig. 2.21 Femtosecond (broadband) radiation can be used for efficient two-photon coupling of narrow atomic transitions [25, 27]

It is interesting to note that the hydrogen laser pulse arises from the hydrogen Balmer alpha line (see Fig. 2.1, *left*), which is at 656.3 nm, essentially the same color as that of a red laser pointer.

2.5 Lasing from Atomic Inert Gases in Air

The potential for achieving lasing from minor species may lead to backward lasing from naturally occurring atmospheric inert gases including argon, neon, krypton, and xenon. In particular, the concentration of argon in the air is 0.8%, close to that of water vapor in room air. The use of inert gases, which are atomic, for multiphoton-pumped lasing, circumvents the need for the dissociation step in the lasing schemes that we discuss next.

Since in the case of naturally occurring atomic species the dissociation step toward atomic lasing is not necessary, it is feasible to think about optical pumping using single ultrashort (femtosecond) pulses. These pulses allow ultrahigh intensities that are required for multiphoton transitions, without the need for very high pulse energies. A good example of a noble gas with two-photon transitions, similar to the ones discussed above in oxygen, nitrogen, and hydrogen, is xenon. Here pumping with 224.29 nm leads to lasing at 834.7 nm, close to the 845 nm lasing in oxygen [26]. To investigate lasing in xenon, we pump it with femtosecond laser pulses at 225 nm that are derived from 50 fs Spectra-Physics Solstice Ti:Sapphire laser at 800 nm, that is, frequency up-converted through optical parametric amplification and nonlinear frequency mixing. Due to the complementarity of the two-photon energy addition of low- and high-frequency spectral components of the femtosecond pump pulse, as illustrated in Fig. 2.21, femtosecond excitation can efficiently couple the two-photon atomic resonance, even though the resonance is much narrower than the bandwidth of the pump radiation [25, 27].

Figure 2.22a illustrates two-photon pumping at 224.29 nm followed by the emission at 834.7 nm in xenon. As shown in Fig. 2.22b, forward- and backward-propagating stimulated emissions are generated simultaneously, within the accuracy of the detection system, which is on the order of 20 ps. Since xenon concentrations in the natural air are less than 100 parts per billion, it is not a strong candidate for an air laser. Nonetheless, lasing from 10 parts per million concentrations in air has been achieved (Fig. 2.23).

The best atomic candidate for generating remote lasing in the atmosphere is argon, because with about 1% concentration, it is the next most abundant air specie after nitrogen and oxygen. However, in order to excite argon atoms in air, three-photon excitation is required due to the ultraviolet absorption edge near 200 nm.

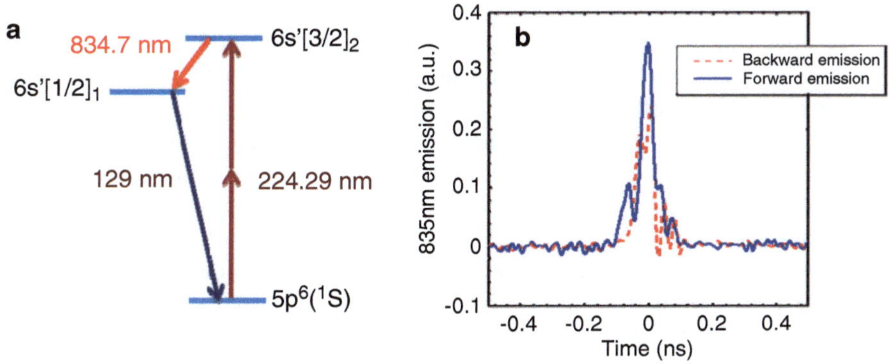

Fig. 2.22 (**a**) Broadband two-photon excitation of Xe leads to a narrowband emission. (**b**) The resulting forward and backward picosecond emissions at 834.7 nm

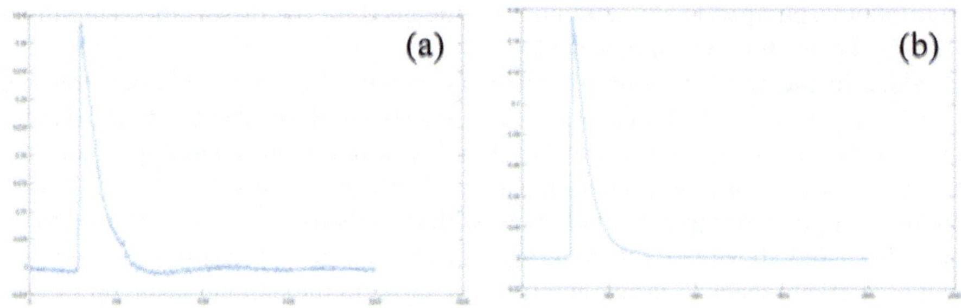

Fig. 2.23 (**a**) Backward-propagating and (**b**) forward-propagating lasing from 10 parts per million of Xe in air

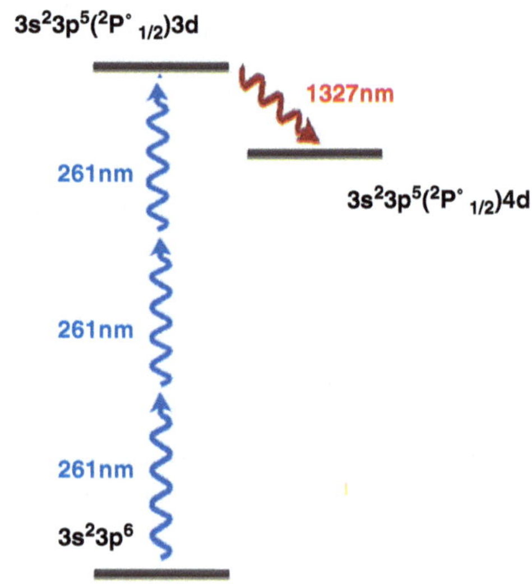

Fig. 2.24 Energy-level diagram of argon showing three-photon excitation by the pump light at 261 nm followed by the emission at 1327 nm

As shown in Fig. 2.24, three-photon pumping at 261 nm brings argon atoms from the ground state $3s^23p^6$ to the upper lasing state $3s^23p^5(^2P^\circ{}_{1/2})3d$. This three-photon excitation is followed by lasing emission from the 3d state to the lower laser state $3s^23p^5(^2P^\circ{}_{1/2})4p$. Backward lasing emission in argon using three-photon excitation has been demonstrated with both 100 ps and 100 fs DUV pump pulses, showing that the bandwidth argument from Fig. 2.21 extends beyond two-photon transitions [14, 15].

Figure 2.25a shows that the pulses emitted in both directions are simultaneous and have a pulse width of 50 ps (FWHM). Although the spectra shown in Fig. 2.25b are limited by the resolution of our spectrometer, they show a 1327 nm emission line much narrower than the femtosecond pump at 261nm.

While three-photon excitation of argon leads to efficient backwards lasing, adding atmospheric air affects the efficiency. As shown in Fig. 2.26, while lasing

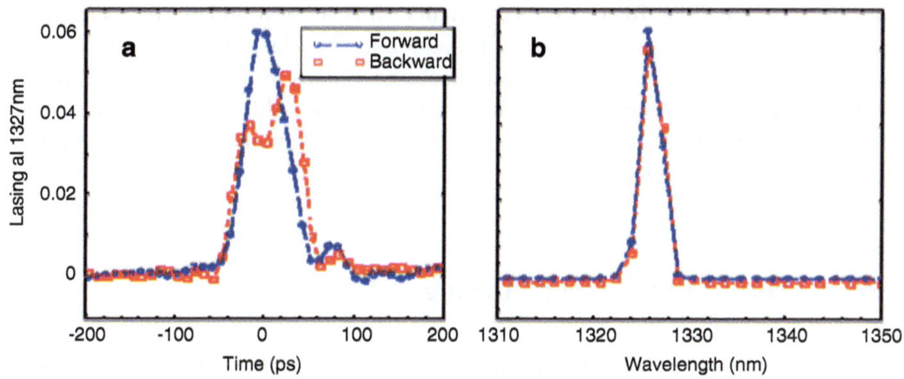

Fig. 2.25 (**a**) Temporal and (**b**) spectral shapes of the 1327 nm emission from Ar in the forward (*squares*) and backward (*circles*) directions

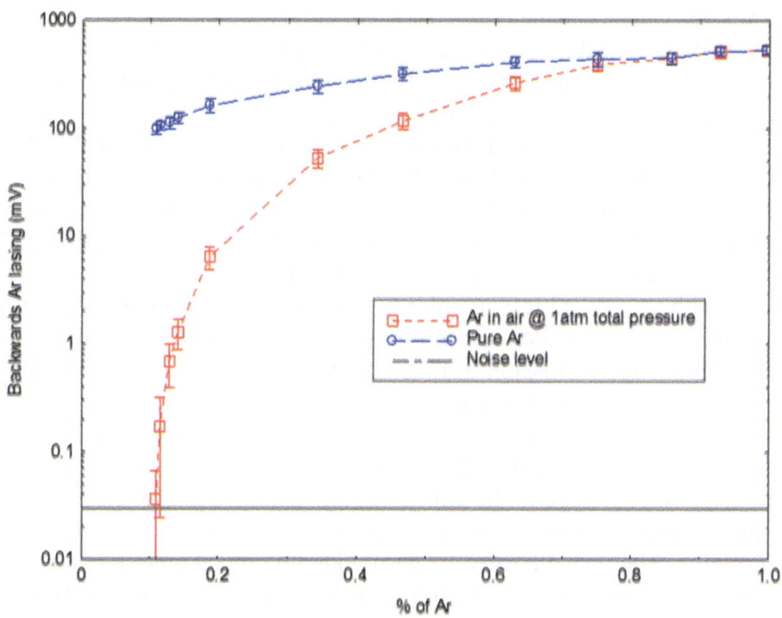

Fig. 2.26 Argon lasing in pure argon and in an argon-air mixture. Lasing is detected with argon concentrations down to 10%

in pure Ar is still strong even at low densities, the air-Ar mixture shows less stimulated emission for the same Ar density. Fig. 2.26 shows detectable backwards emission in atmospheric air containing as low as 10% Ar. The higher lasing threshold in atmospheric air can be attributed to the rapid collisional transfer of energy from the excited Ar states to the electronic excited energy levels of nitrogen and oxygen. The goal of obtaining backwards lasing from Ar in atmospheric air can be reached by further improvements on detection and by increasing the pump energy to allow for longer gain paths and thus lasing at lower Ar partial pressure. For this purpose using femtosecond pulses is beneficial, because it avoids the potential problem of avalanche ionization. Additionally, femtosecond pumping occurs faster than the collision time, so collisional deactivation of argon during pumping does not occur.

2.6 "Around-the-Corner" Lasing

Since the outcome of an optical gain is exponential in the length of the active medium, the direction of lasing is highly dependent on the geometry of the gain volume. In the cases of backward and forward lasing, discussed above, the gain volume was elongated along the propagation direction of the pump beam. However, through the modification of the geometry of the gain volume, the direction of the lasing emission can be changed from being either forward or backward propagating. That opens up the possibility of an "around-the-corner" air laser.

In the "around-the-corner" laser, a cylindrical, instead of a spherical, lens focuses the UV pump beam, creating a gain volume elongated in the direction perpendicular to the direction of propagation of the pump beam. Such an arrangement is shown in Fig. 2.27. In the experimental demonstration of this approach, the dissociation of nitrogen is driven by a Nd:YAG laser. The dissociating pre-pulse is followed, with a 100 ns delay, by a 100 ps UV pump pulse, which is focused with a cylindrical lens. The dissociating nanosecond laser is focused into the same region as the 100 ps UV pulses, but with a spherical lens in order to build up sufficient intensity for efficient dissociation. Fast photodiodes placed on the either side of the elongated gain volume detect the 745 nm emission from atomic nitrogen through bandpass filters. In addition, the emission pattern of the sideways lasing is visualized by photographing its projection on a screen with a CCD camera. As shown in Fig. 2.28, a wide band of light at 745 nm is emitted from the focal region sideways.

The 1.5 by 10 cm image of the emission pattern shown in Fig. 2.28a is obtained by averaging, over 100 laser shots, the image on a screen placed 10 cm away from the gain region. The image clearly shows a wide horizontal emission band. This is to be expected because the focal region is close to be rectangular. On the contrary,

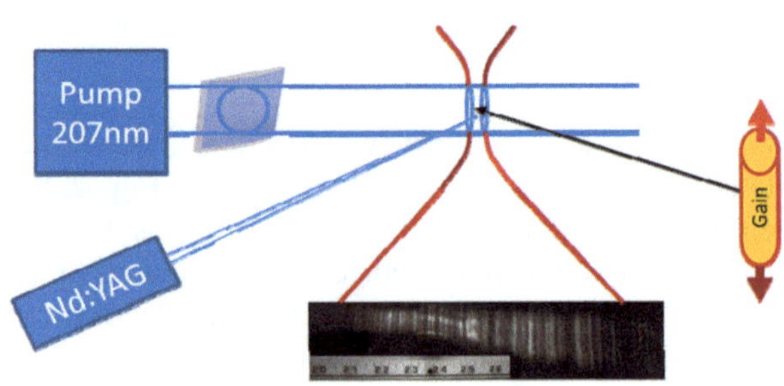

Fig. 2.27 Experimental setup for air lasing at 90° angle with respect to the propagation direction of the pump beam. An elongated gain region is produced through the focusing of the pump beam with a cylindrical lens

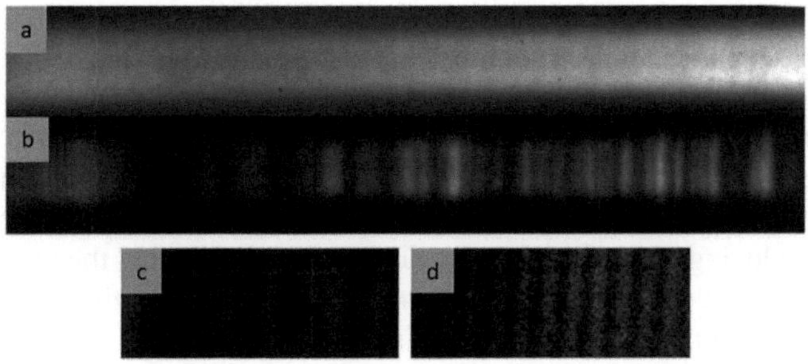

Fig. 2.28 (**a**) Multi-pulse averaged and (**b**) single-shot radiation patterns emitted by the atomic nitrogen laser at the 90° angle with respect to the direction of the pump beam. The pump is focused with a cylindrical lens, as illustrated in Fig. 2.27c, d. Blow-ups of the emission pattern showing vertical lines with various perioducity, which indicates a high degree of coherence of the sideways lasing

single-shot images, such as the one shown in Fig. 2.28b, show that the emission pattern consists of a series of narrow vertical lines, the positions of which randomly change from one laser shot to another. These lines appear according to the fluctuating distribution of excited nitrogen atoms. The emission is strongest along the directions along which higher numbers of emitters are aligned. Although the longest dimension of the gain volume is along the direction perpendicular to the direction of propagation of the laser beam, the gain region is also extended over the Rayleigh range of the cylindrical focusing optic. There are various paths along which stimulated emission can take place. Since stimulated emission is coherent, these spatial modes interfere with each other, creating a series of vertical lines. The emission patterns are similar to the either sides of the gain volume.

Successive blowups of the emission pattern shown in Fig. 2.28c, d demonstrate high contrast of the interference, indicating a high degree of coherence of the lasing. Since the emission mechanism is the same as that in the backward air laser, the coherence time of the emission is on the order of several tens of picoseconds, which matches the optical path length along the gain region. In principle, higher spatial coherence can be achieved by arranging a spatially localized series of gain regions that would sequentially amplify stimulated emission in a particular direction.

2.7 Conclusions

Backward-propagating laser radiations from two-photon pumped atomic oxygen, nitrogen, and hydrogen, which are produced by dissociation of molecules in room air, possess similar characteristics: high spatial and temporal coherence and close to transform limited properties with pulse durations between 10 and 30 ps, independent of the pulse length of the pumping laser. These laser beams are generated over millimeter scale paths that exhibit optical gain values as high as ~60/cm. The exponential dependence of laser emission on the path length and the atom number density, as well as the lack of sensitivity to the dephasing collision rates, are indicative of stimulated emission being the dominant emission mechanism. The potential for developing a backward laser based on the naturally occurring inert gases such as argon is very attractive as it eliminates the need for the dissociation step of the excitation, reducing pulse-to-pulse fluctuations and enhancing the potential for the use of various modulation methods for trace detection. Changing the geometry of the gain volume through different focusing of the pump beam provides the opportunity for steering air lasing in the directions other than forward and backward.

Acknowledgments The authors acknowledge the support from the Office of Naval Research and from NASA through the SBIR program to MetroLaser, Inc.

References

1. R.B. Miles, A. Dogariu, J.B. Michael, Bringing bombs to light. IEEE Spectr. **49**, 38 (2012)
2. A. Dogariu, J.B. Michael, R.B. Miles, Remote air lasing for trace detection. Proc. SPIE **8024**, 80240H (2011)
3. A. Dogariu, J.B. Michael, M.O. Scully, R.B. Miles, High gain backward lasing in air. Science **331**, 442 (2011)
4. A. Dogariu, J. B. Michael, M. O. Scully, and R. B. Miles, Remote backwards lasing in air. Conference on Lasers and Electro-Optics CLEO'2011 (Baltimore, MD 2011), paper JTuE2
5. A. Dogariu, J. Michael, and R. Miles, High gain atomic oxygen lasing in air. 42nd AIAA Plasmadynamics and Lasers Conference (Honolulu, HI, 2011), paper 2011–4001
6. A. Dogariu, J.B. Michael, R.B. Miles, Standoff stimulated emission in air. Proc. SPIE. **8366**, 20 (2012)
7. A. Dogariu and R. B. Miles, Lasing in atmospheric air: similarities and differences of oxygen and nitrogen. in *Frontiers in Optics 2013, OSA Technical Digest* (2013), paper LTh2H.2
8. A. Dogariu and R. B. Miles, Nitrogen lasing in air. Conference on Lasers and Electro-Optics CLEO'2013 (San Jose, CA, 2013), paper QW1E.1
9. A. Dogariu, R. B. Miles, Backwards nitrogen double lasing in air for remote trace detection. in *Imaging and Applied Optics 2014, OSA Technical Digest* (2014), paper LW2D.3
10. A. Laurain, M. Scheller, P. Polynkin, Low-threshold bidirectional air lasing. Phys. Rev. Lett. **113**, 253901 (2014)
11. A. Dogariu and R. Miles, Remote backward-propagating lasing of nitrogen and oxygen in air. Conference on Lasers and Electro-Optics CLEO'2015 (San Jose, CA, 2015), invited paper SM1N.1
12. A. Dogariu, T.L. Chng, and R. B. Miles, Backwards lasing from minor species in air. 46th Winter Colloquium on the Physics of Quantum Electronics, (Snowbird, UT, 2016)
13. A. Dogariu, T. L. Chng, and R. B. Miles, Remote backward-propagating water lasing in atmospheric air. in *CLEO: 2016, OSA Technical Digest* (2016), paper AW4K.5
14. A. Dogariu, J. Li, R. B. Miles, Three-photon pumped backwards lasing in argon. OSA Light (Energy and Environment Congress, Suzhou, 2015)
15. A. Dogariu, R.B. Miles, Three-photon femtosecond pumped backwards lasing in argon. Opt. Express **24**, A544 (2016)
16. P.R. Hemmer, R.B. Miles, P. Polynkin, T. Siebert, A.V. Sokolov, P. Sprangle, M.O. Scully, Standoff spectroscopy via remote generation of a backward-propagating laser beam. Proc. Nat. Acad. Sci. **108**, 3130–3143 (2011)

17. A. E. Siegman, *Lasers*. (Mill Valley, University Science Books, 1986).
18. B.A. Reagan, K.A. Wernsing, A.H. Curtis, F.J. Furch, B.M. Luther, D. Patel, C.S. Menoni, J.J. Rocca, Demonstration of a 100-Hz repetition rate gain-saturated diode-pumped table-top soft x-ray laser. Opt. Lett. **37**, 3624 (2012)
19. J.C. Macgillivray, M.S. Feld, Superradiance in atoms and molecules. Contemp. Phys. **22**, 299 (1981)
20. M.N. Schneider, R.B. Miles, Microwave diagnostics of small plasma objects. J. Appl. Phys. **98**, 0033301 (2006)
21. T.A. Cool, Quantitative measurement of NO density by resonance three-photon ionization. Appl. Opt. **23**, 1559 (1984)
22. A. Dogariu, R.B. Miles, Detecting localized trace species using radar REMPI. Appl. Opt. **50**, A68 (2011)
23. D.C. Dai, Brief comment: Dicke superradiance and superfluorescence find application for remote sensing in air. ArXiv. **1108**, 5360 (2011)
24. A.J. Traverso, R. Sanchez-Gonzalez, L. Yuan, K. Wang, D.V. Voronine, A.M. Zheltikov, Y. Rostovtsev, V.A. Sautenkov, A.V. Sokolov, S.W. North, M.O. Scully, Coherence brightened laser source for atmospheric remote sensing. Proc. Nat. Acad. Sci. **109**, 15185 (2012)
25. S. Alisauskas, A. Baltuska, R. Boyd, P. Polynkin, Backward air lasing with femtosecond pumping. CLEO Europe 2015, postdeadline paper PD-A.4
26. A. Dogariu, T. L. Chng, and R. Miles, Towards remote magnetic anomaly detection using Radar REMPI. CLEO 2014, paper SM4E.4
27. W.D. Kulatilaka, J.R. Gord, V.R. Katta, S. Roy, Photolytic-interference-free, femtosecond two-photon fluorescence imaging of atomic hydrogen. Opt. Lett. **37**, 3051 (2012)

Chapter 3
The Role of Electron Collisions in Lasing in Neutral and Singly Ionized Molecular Nitrogen

Yi Liu, Pengji Ding, Aurélien Houard, and André Mysyrowicz

3.1 Introduction

In this chapter, we will discuss lasing actions in the air that follow the excitation with a short intense laser pulse at 800 nm. We will successively analyze two types of laser actions. The first type is based on the optical transition between the excited triplet states of the neutral nitrogen molecule. Based on the study of the dependence of the lasing signal on the polarization ellipticity of the pump pulse, we unambiguously attribute gain mechanism in this scheme to the electron collisions with neutral nitrogen molecules that result in population inversion. Experimental results on the dynamics of emissions in the forward and backward directions with respect to the direction of the pump pulse are confirmed by numerical simulations based on the Maxwell-Bloch equations. The second type of lasing stems from the transition between the second electronically excited state and the ground state of a singly ionized nitrogen molecule. After reviewing current interpretations of this emission process, which remains to be a controversial issue, we will focus on our interpretation

Y. Liu (✉)
Laboratorie d'Optique Appliquée, ENSTA ParisTech, CNRS, Ecole Polytechnique, Université Paris-Saclay, 828 Boulevard des Maréchaux, 91762 Palaiseau, France

Shanghai Key Lab for Modern Optical System, University of Shanghai for Science and Technology, 200093 Shanghai, China
e-mail: yi.liu@ensta-paristech.fr

P. Ding
Laboratorie d'Optique Appliquée, ENSTA ParisTech, CNRS, Ecole Polytechnique, Université Paris-Saclay, 828 Boulevard des Maréchaux, 91762 Palaiseau, France

School of Nuclear Science and Technology, Lanzhou University, 730000 Lanzhou, China

A. Houard • A. Mysyrowicz
Laboratorie d'Optique Appliquée, ENSTA ParisTech, CNRS, Ecole Polytechnique, Université Paris-Saclay, 828 Boulevard des Maréchaux, 91762 Palaiseau, France
e-mail: andre-mysyrowicz@ensta-paristech.fr

P. Polynkin, Y. Cheng (eds.), *Air Lasing*, Springer Series in Optical Sciences 208, https://doi.org/10.1007/978-3-319-65220-7_3

that links stimulated emission in this scheme to superradiance. We will argue that electron recollisions play an essential role in establishing the superradiant gain.

There is currently a strong interest to develop an "air laser" in free space. Such a laser uses the components of air or their ions as a gain medium [1–23]. In principle, such a cavity-free laser could be realized at large distances by exploiting the process of laser filamentation [24] that can be used to pump such laser actions. Filamentation offers the possibility to maintain very high laser intensities over multimeter paths and can be initiated at kilometer-scale standoff distances. It has been shown to turn air into an active medium with high gain, which is a prerequisite for cavity-free lasing. In principle, both forward-propagating and backward-propagating lasing schemes are of interest. The remote generation of a coherent beam at a different wavelength from that of the pump beam, especially the wavelength in the UV, could be very useful for various atmospheric studies. In the case of forward-propagating air lasing, one can envision the setup where pump pulses are emitted from satellites or floating balloons, and the generated air lasing signals carry information about the atmosphere toward an Earth-based receiver. With a backward-propagating laser pulse emanating from a remote location in the sky, coherent optical detection methods such as stimulated Raman scattering can be used for remote sensing in the air using a pump pulse launched from the Earth [13]. In principle, this approach has the potential to increase, by orders of magnitude, the detection sensitivity compared to the traditional incoherent detection schemes of optical remote sensing, although there has been no demonstration of this potential yet. At present, the main task is to identify the physical mechanisms that are responsible for air lasing. This chapter is devoted to this aspect of the problem and is restricted to the study of air lasing obtained using short pump pulses at 800 nm from Ti/sapphire chirped-pulse amplified (CPA) laser systems.

The concept of air lasing has been put forth over a decade ago. The last 5 years have witnessed significant progress toward its realization [1–3, 9, 12]. So far, three different types of schemes for air lasing have been demonstrated. In the first scheme, picosecond or nanosecond ultraviolet (UV) pulses (226 nm) are used to dissociate oxygen and nitrogen molecules in the ambient air and subsequently pump the resulting atomic fragments through two-photon resonant processes [1]. In the case of oxygen, both backward- and forward-propagating stimulated emissions at 845 nm have been observed in the experiments, with the backward-propagating lasing energy as high as 1 microjoule [8]. However, the application of this scheme for remote sensing is limited by high absorption and scattering of the UV pump pulses in the atmosphere. This limitation motivates search for alternative schemes based on optical pumping in the visible or IR.

One such alternative is based on the creation of population inversion in neutral nitrogen molecules. Backward-stimulated emission from neutral nitrogen molecules inside a laser plasma filament was first suggested in 2003, based on the observed exponential growth of the backward UV emission with the length of the plasma filament [12]. In 2012, D. Kartashov and coworkers focused a mid-infrared femtosecond laser pulses (at either 3.9 or 1.03 μm) inside a high-pressure mixture of argon and nitrogen gas. They observed backward-propagating stimulated

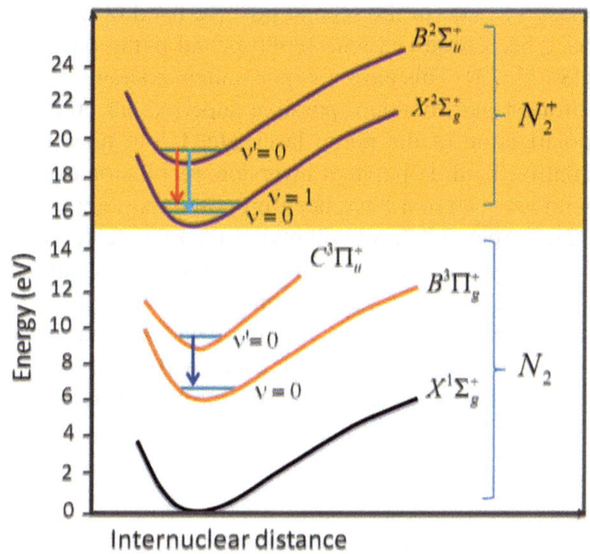

Fig. 3.1 Schematic energy diagram of the neutral and singly ionized nitrogen molecules. Historically, the transition $C^3\Pi_u^+ \rightarrow B^3\Pi_g^+$ in neutral N_2 is termed the second positive band of nitrogen, while the transition $B^2\Sigma_u^+ \rightarrow X^2\Sigma_g^+$ in N_2^+ is termed the first negative band of nitrogen

emissions at 337.1 and 357 nm at the optimal argon gas pressure of 5 bars and nitrogen pressure of 2 bars [9]. The emission at 337.1 or 357 nm has been identified as being due to the transition between the third and second excited triplet states of neutral nitrogen molecules, i.e., $C^3\Pi_u \rightarrow B^3\Pi_g$. The relevant energy level diagram of neutral and ionic nitrogen molecules is presented in Fig. 3.1. The population inversion mechanism between the $C^3\Pi_u$ and $B^3\Pi_g$ states is attributed to the traditional Bennet mechanism, where collisions transfer the excitation energy of argon atoms to molecular nitrogen. Unfortunately, this method cannot be applied for remote generation of air lasing emission because of its requirement of high-pressure argon gas ($p > 3$ bar). In another work, researchers used a 10 ps, 10 J laser to pump ambient air and observed forward 337 nm stimulated based on the population inversion between the $C^3\Pi_u$ and $B^3\Pi_g$. However, no evidence of backward lasing was observed [14].

Recently, it has been demonstrated that the filament plasma column created by an intense circularly polarized femtosecond laser pulse at 800 nm in pure nitrogen or in the nitrogen-oxygen mixture can act as an optical amplifier and give rise to lasing at 337.1 nm wavelength in both backward and forward directions [2, 16, 19]. This simple method for lasing of neutral nitrogen molecules with the widely available 800 nm femtosecond pumping holds great potential for future applications. In Sect. 3.2, we will present experimental and simulation results based on this scheme and discuss its underlying physical mechanism. The role of electron collisional excitation in the creation of population inversion will be discussed in detail.

A different scheme for air lasing has been reported in 2011 by a research group from the Shanghai Institute of Optics and Fine Mechanics (SIOM) [2]. In this scheme, the lasing manifests itself as a strong, forward-propagating, narrow-bandwidth emission from the $C^3\Pi_u^+ \rightarrow B^3\Pi_g^+$ transition in singly ionized nitrogen ions N_2^+, i.e., in the second positive band of nitrogen (Fig. 3.1). These emissions have been observed with femtosecond pump pulses in the UV, visible and mid-IR [19, 22, 23]. This emission phenomenon shows several unusual features, including ultrafast gain buildup, possible superradiant emission, strong dependence on the polarization of the pump laser, etc. Up to now, its physical mechanism remains controversial. Population inversion, recollision excitation, superradiance, and laser without inversion have been proposed as potential gain mechanisms [2, 4–6, 21, 22]. In Sect. 3.2.2, we will review the main features of this lasing effect with pumping at 800 nm. We will further discuss various gain mechanisms proposed in recent literature. We will give our interpretation of the origin of this lasing in terms of superradiance, based on the time-resolved measurements of the emission. Finally, we will present experimental results with pump pulses of varying polarization ellipticity and suggest electron recollision excitation as a potential excitation mechanism responsible for the stimulated emission. We will present experimental results that support that hypothesis.

3.2 Lasing in Neutral Nitrogen in a Femtosecond Laser Filament: The Gain Mechanism and Temporal Dynamics

Amplified spontaneous emission (ASE) at 337.1 nm in both backward and forward directions can be obtained from laser filaments in the nitrogen gas or in the nitrogen-oxygen mixture under pumping with circularly polarized 800 nm femtosecond laser pulses [3]. In Sect. 3.2.1, we will present the main features of this lasing scheme. In Sect. 3.2.2, the underlying mechanism will be discussed in detail, and the electron collisional excitation for population inversion will be confirmed both experimentally and through calculations. In Sect. 3.2.3, time-resolved measurements of the emission in the ASE and seeded regime will be presented. Finally, a theoretical model based on the Maxwell-Bloch equations will be presented, and the resulting simulations will be compared with experiments.

3.2.1 Main Features of N_2 Lasing with Circularly Polarized Femtosecond Pumping

The experimental setup for the studies of lasing in N_2 is shown in Fig. 3.2. Femtosecond laser pulses with duration of 45 fs are focused by a convex lens of 1000 or 500 mm in a gas cell filled with 1 bar of pure nitrogen gas or a mixture of nitrogen and oxygen. A broadband dielectric beam splitter is used to direct the incident pump beam into the gas chamber while routing the backward-propagating UV emission from the gas plasma to the detector. A quarter-wave plate is placed in front of the input window of the gas chamber, to change the laser polarization from linear

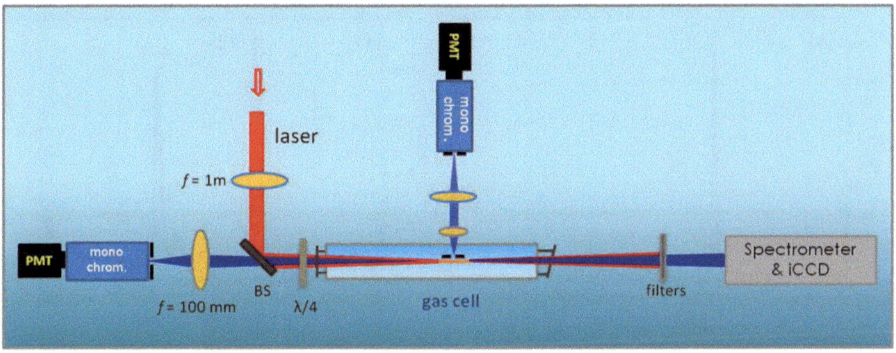

Fig. 3.2 Schematic of the experimental setup for the measurement of backward-propagating and forward-propagating lasing and transverse fluorescence in filaments

to circular. The backward-propagating emission is focused by an $f = 100$ mm fused silica lens to the slit of a monochromator (Jobin-Yvon H-20 UV, grating: 1200 g/mm) combined with a photomultiplier tube (PMT). The forward-propagating emission is analyzed by a spectrometer (Ocean Optics, UV 4000) or by an iCCD camera. The transverse fluorescence from the plasma channel is also measured with a monochromator and a PMT.

We have measured the following parameters of both backward-propagating and forward-propagating emissions from filaments in pure nitrogen and in the ambient air: emission spectrum, spatial profile, and dependence on the pump pulse energy and polarization. Below we present our main results.

3.2.1.1 Spectrum Analysis of the Backward Emission in Pure Nitrogen Gas

In Fig. 3.3, the spectra of the backward UV emission are shown for both circularly and linearly polarized pump pulses at 800 nm. The emission intensity at 337 nm in the case of circular laser polarization is about 40 times larger than that of linear laser polarization. For the other lines at 315, 357, 380, and 405 nm, an increase of signal by a factor of ~1.5 is observed when the laser pulse polarization is changed from linear to circular. This remarkable behavior of the 337 nm signal suggests that backward-stimulated emission is initiated with circularly polarized laser pulses.

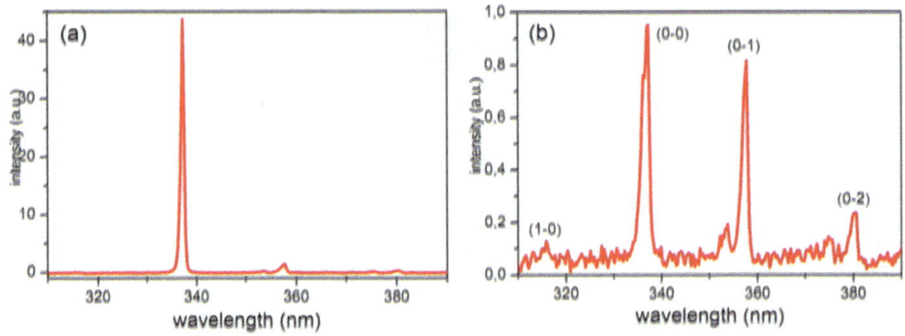

Fig. 3.3 Spectrum of the backward emission with circularly (**a**) and linearly (**b**) polarized pump pulses. The numbers in parenthesis in (**b**) designate the vibration quantum number of the initial and final states of the transition

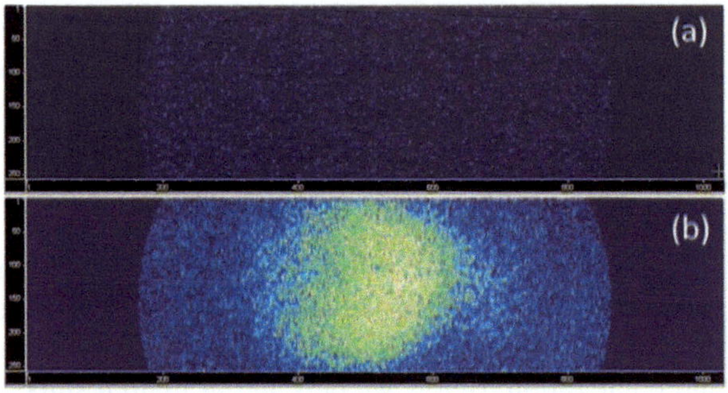

Fig. 3.4 Backward emission at 337 nm from plasma filaments observed with linearly (**a**) and circularly (**b**) polarized pump pulses. The pump pulse energy is about 10 mJ and the nitrogen pressure is 1 bar

3.2.1.2 Spatial Profile of the Backward Emission

We have measured the spatial profile of the backward emission at 337 nm with an ICCD camera (Princeton Instruments, PI-MAX). An interference filter centered at 337 nm with a bandwidth of 10 nm was installed before the CCD camera to block any backward-propagating optical background at the wavelengths other than that of the lasing. The results are shown in Fig. 3.4, for both linearly and circularly polarized pump pulses. A well-defined beam is observed when circularly polarized pump is used, while no lasing signal is detected with linearly polarized pump beam, as shown in Fig. 3.4a, b respectively. This observation unambiguously confirms the occurrence of backward-propagating stimulated emission at 337 nm in the filaments produced in nitrogen gas by circularly polarized femtosecond laser pulses.

Analysis of polarization properties of this backward emission reveals that the emission is not polarized [3], which indicates that it is an unseeded amplified spontaneous emission (ASE).

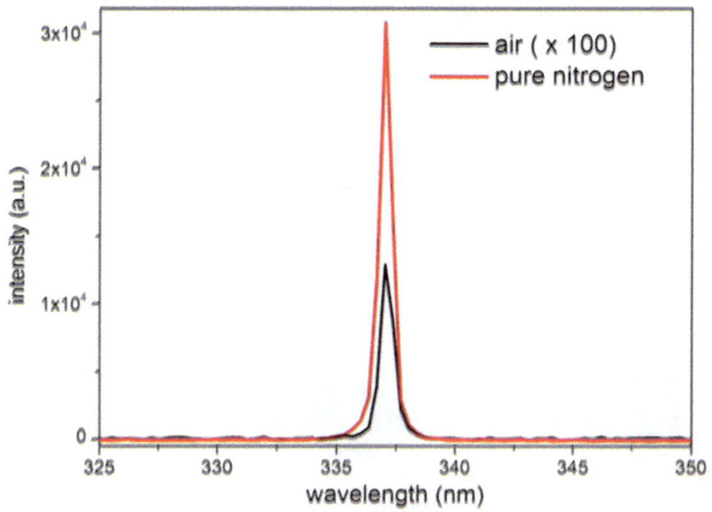

Fig. 3.5 Forward lasing spectrum in ambient air and pure nitrogen. The signal obtained in the air has been multiplied by a factor of 100

3.2.1.3 Lasing Action in Ambient Air

In the experiments in pure nitrogen gas, we have noticed that forward-propagating emission at 337 nm is also generated, and its strength is much higher than that of the backward-propagating emission. Typically, the forward-propagating 337 nm lasing intensity is 3 orders of magnitude more intense than the intensity of the backward-propagating ASE. Very recently, we have demonstrated that 337 nm lasing can be generated in the ambient air in the forward direction [19], when it is pumped by circularly polarized 800 nm pulses, although the intensity of the lasing is 250 times smaller than that in pure nitrogen (Fig. 3.5). This observation suggests that oxygen acts as an efficient quenching agent for the 337 nm nitrogen lasing [3]. To achieve backward 337 nm emission in the ambient air, a higher optical gain or a longer effective gain length is necessary, which may be possible with higher energies of the pump pulses than what has been used previously.

3.2.1.4 Amplification of an External Seed Pulse

In the presence of population inversion, an externally injected seed pulse at the wavelength within the gain bandwidth is expected to be amplified. We have conducted experiments on the amplification of an external seed signal propagating in both the same and opposite directions with respect to the direction of the pump pulse. In both cases, the seed pulse can be amplified by two orders of magnitude. In Fig. 3.6, we present the results for the amplification of the backward-propagating seed pulse, observed in pure nitrogen. In Fig. 3.6a, we show the spatial profile of the backward

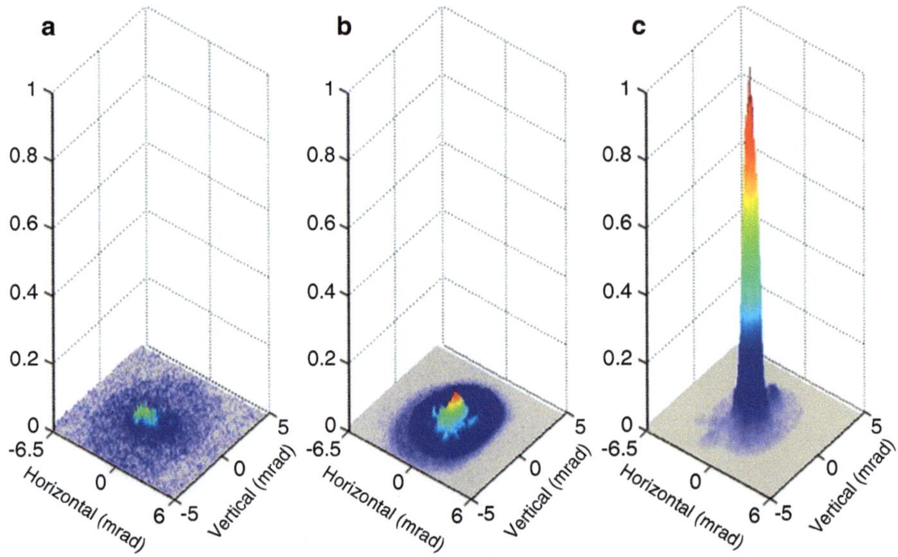

Fig. 3.6 Spatial profile of the backward ASE (**a**), the seed pulse (**b**), and the amplified 337 nm radiation (**c**). The range of angles in each panel is 12.5 mrad × 10 mrad

ASE. The ASE has Gaussian intensity distribution with angular divergence of 9.2 mrad. The spatial profile of the seed pulse is shown in Fig. 3.6b. In the presence of both pump and seed pulses, an intense 337 nm radiation is generated, as shown in Fig. 3.6c. The amplified stimulated emission has a divergence angle of ~3.8 mrad, much smaller than that of both the ASE and of the seed pulse.

3.2.2 Discussion of the Mechanisms for the Creation of Population Inversion

In the above section, we have shown that lasing action at 337 nm in filaments can only be achieved with circularly polarized 800 nm femtosecond pulses. Experimentally, we have verified that a slight deviation of the laser polarization from the circular polarization state leads to an extremely steep drop of the lasing signal [3].

How should we understand this very strong dependence of the 337 nm lasing on the polarization of the pump? Direct optical strong-field excitation of the triplet state $C^3\Pi_u^+$ of nitrogen is a spin-forbidden process. Two indirect excitation mechanisms have been proposed. The first scheme involves the dissociative recombination through the reactions $N_2^+ + N_2 + N_2 \rightarrow N_4^+ + N_2$ followed by $N_4^+ + e \rightarrow N_2\left(C^3\Pi_u^+\right)$ [25]. Another recently proposed scenario suggests that the collision-assisted intersystem crossing from the excited singlet state is the dominant process that populates the triplet state, while the dissociative recombination is a minor contributor [26].

Contrary to our experimental observations, both processes should favor linearly polarized pump for the following reasons. In the case of dissociative recombination, density of the N_2 molecules in the $C^3\Pi_u^+$ state depends on the density of N_2^+, which is more efficiently produced by linearly polarized radiation in the intensity range 10^{13}–10^{15} W/cm^2 [27]. With the intersystem crossing mechanism, similar dependence on the laser polarization is expected, because the transition from the fundamental singlet state of N_2 to an intermediate singlet state is more efficient with linearly polarized laser pulses. The question is why does circular but not linear polarization of the pump laser pulses enable the 337 nm lasing?

There is an important difference between electron dynamics in the cases of strong-field ionization by linearly and circularly polarized laser fields [28, 29]. With linearly polarized laser pulses, free electrons are left with low kinetic energies at the end of the pump pulse because they are alternately accelerated and decelerated by the laser field during every optical cycle. With a circularly polarized laser field, electrons are always accelerated away from their parent molecular ion. At the end of the laser pulse, they acquire average energy of ~$2U_p$, where $U_p = e^2 I / 2c\varepsilon_0 m_e \omega_0^2$ is the ponderomotive potential of the electron in a linearly polarized laser field and ε_0, m_e, I, and ω_0 are the vacuum permittivity, electron mass, optical intensity, and frequency of the laser field, respectively [16]. For laser intensity of $I = 1.4 \times 10^{14}$ W/cm^2, which is a representative intensity level inside air filaments under tight external focusing conditions [30, 31], a large number of electrons with kinetic energies around $2U_p \geq 16$ eV can be produced. Such an energy is sufficient for the excitation of nitrogen molecules from their ground state to the excited triplet state through the following inelastic collision reaction:

$$N_2\left(X^1\Sigma_g^+\right) + e = N_2\left(C^3\Pi_u^+\right) + e \tag{3.1}$$

This reaction is the dominant mechanism responsible for the establishment of population inversion between the $C^3\Pi_u^+$ and $B^3\Pi_g^+$ states of nitrogen in the conventional discharge-pumped nitrogen laser, where electrons are accelerated to obtain sufficient energy by the discharge electric field [32]. Therefore, the effectiveness of circular laser polarization in creating population inversion stems from the fact that the photoelectrons generated in a circularly polarized laser field are left with a substantial kinetic energy immediately after the passage of the pump laser pulse.

To further confirm the role of collisional excitation in the creation of population inversion, we have calculated the electron kinetic energy distribution for different polarization ellipticities of the pump field. The transverse kinetic energy distribution for the photoelectrons can be derived by using semi-analytical laws. The integration of Newton's equations of motion for the electron leads to the transverse momentum $\vec{p}(t) = -e\left(\vec{A}(t) - \vec{A}_0\right)$, where $\vec{A}(t)$ is the vector potential of the optical field at the time t and $\vec{A}_0$ is the vector potential at the instant when the electron is liberated, when it is assumed to have zero kinetic energy. After the passage of the pulse, $\vec{A}(t)$ vanishes. The transverse momentum becomes $p(\infty) = -eA_0$, and the transverse kinetic energy reads $E_{kin}(t_0) = e^2 A_0^2 / 2m$, where t_0 is the time instant, within the pulse, when the electron was liberated. We can infer the vector potential by integrating the equation $\vec{E} = -\partial\vec{A}/\partial t$, using the analytical form for the electric field with a cosine envelope

$$E = \begin{cases} E_0 \cos(\pi t/T)\begin{bmatrix} \cos(\omega_0 t+\theta)\vec{u}_x \\ +\varepsilon\sin(\omega_0 t+\theta)\vec{u}_y \end{bmatrix}, & \text{for} -T/2 < t < T/2 \\ 0, & \text{for } t - T/2 \text{ and } t > T/2, \end{cases} \tag{3.2}$$

where θ is an arbitrary carrier-envelope phase. The kinetic energy of the electron, born at the time t_0, after the acceleration by the pulse reads:

$$E_{kin} = 2U_p\left(\pi t_0/T\right)\left[1-\left(1-\varepsilon^2\right)\cos^2\left(\omega_0 t_0+\theta\right)\right] \tag{3.3}$$

Therefore, all electrons generated between t_0 and $t_0 + \Delta t$, with probability $\left(\partial n_e/\partial t\right)_{t_0} \times \left(\Delta t/n_{e,\infty}\right)$, where $n_e(t)$ is the electron density calculated from the rate equations and $n_{e,\infty}$ is the total electron density produced by the pulse, will have kinetic energies between $E_{kin}(t_0)$ and $E_{kin}(t_0 + \Delta t)$. From the above calculation, the maximum kinetic energy of the electron is $2U_p$ for a circularly polarized pulse ($\epsilon = 1$), when the electron is born at the peak of the field envelope (at $t = 0$).

A parametric representation of the kinetic energy distribution is shown with a continuous curve in Fig. 3.7. In the case of linear laser polarization, the majority

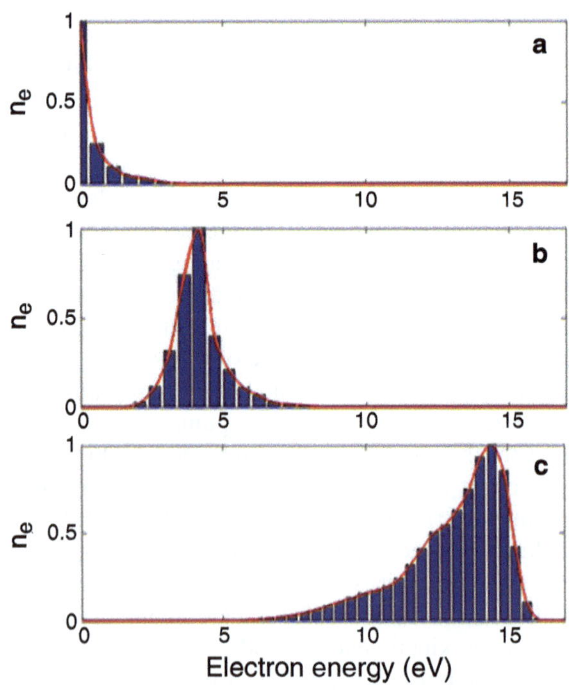

Fig. 3.7 Numerically simulated electron energy distribution in the case of linearly (**a**), elliptically with $\epsilon = 1/2$ (**b**), and circularly (**c**) polarized laser pulses. The laser intensity used in the simulations is 1.4×10^{14} W/cm^2 (Credit: Arnaud Couairon)

of electrons are left with energy below 1 eV, as shown in Fig. 3.7a. By contrast, an almost monoenergetic distribution with the peak around 14.6 eV is produced by circular laser polarization, as shown in Fig. 3.7c. An intermediate distribution is obtained with an elliptically polarized pulse, as shown, for the case of ellipticity $\epsilon = 1/2$, in Fig. 3.7b. Note that the collision process $N_2\left(X^1\Sigma_g^+\right) + e = N_2\left(C^3\Pi_u^+\right) + e$ requires a minimum electron kinetic energy of 14 eV to be effective [32]. This explains the high experimentally observed sensitivity of the nitrogen gain to the ellipticity of the pump.

3.2.3 Time-Resolved Measurements of N_2 Lasing

For time-resolved measurements of gain dynamics in the N_2 lasing, femtosecond laser pulses (800 nm, 45 fs, 100 Hz, 16 mJ) were split into 3 beams (beams *a*, *b*, *c*). Time delays between these pulses were controlled by optical delay lines. For the creation of the nitrogen filament amplifier, a circularly polarized femtosecond pulse of 9 mJ energy (beam *a*) was focused by a convex lens with $f = 1$ m into a chamber filled with nitrogen gas. Both backward and forward ASE at 337.1 nm from the filamentary plasma with a length $l \sim 30$ mm were recorded. To obtain the seeding pulse at 337.1 nm, the second harmonic of the beam *b* was generated in a 1-mm-thick BBO crystal and then focused into a 20-mm-thick fused silica sample. A narrow spectral bandwidth (~10 nm bandwidth) of the emerged broadband supercontinuum was selected with an interference filter centered at 337 nm. The seed pulse could be injected into the plasma in both directions [16].

To characterize the duration of the amplified pulse in the forward direction, we employed a cross correlation technique schematically shown in Fig. 3.8a. The forward-propagating 337.1 nm ASE and the amplified lasing radiation were focused together with the third 800 nm beam *c* on a 2-mm-thick type-I BBO crystal cut at 50.7°, for efficient sum-frequency generation (SFG) of the signal at 238 nm wavelength, as illustrated in Fig. 3.8b. By recording the SFG signal at 238 nm as a function of the delay between the 337.1 nm lasing signal and the weak 800 nm probe pulse *c*, the temporal profile of the 337.1 nm lasing pulse was obtained.

We measured the temporal structure of the forward-propagating 337 nm nitrogen lasing under the conditions of 1 bar pressure, $f = 100$ cm focusing of the pump, and 9 mJ pump pulse energy. The results for the seed pulses centered at 340 nm, amplified pulses at the 337 nm peak lasing wavelength, and the 337 nm ASE are shown in Fig. 3.9a. The 337 nm ASE pulse has pulse duration of 14 ps, and its peak is delayed by 12.5 ps with respect to the pump pulse. By injecting a co-propagating seed pulse with a duration of 1.5 ps, an externally seeded 337 nm lasing pulse is generated with a pulse duration of 3 ps and a 7 ps temporal advancement relative to the 337 nm ASE pulse. Figure 3.9b shows the corresponding simulation results, to be discussed later.

We further applied the cross correlation technique to the measurement of the temporal profile of the backward-propagating externally seeded 337.1 nm lasing pulse. The result is shown in Fig. 3.10a. The radiation peak is delayed by 14 ps with

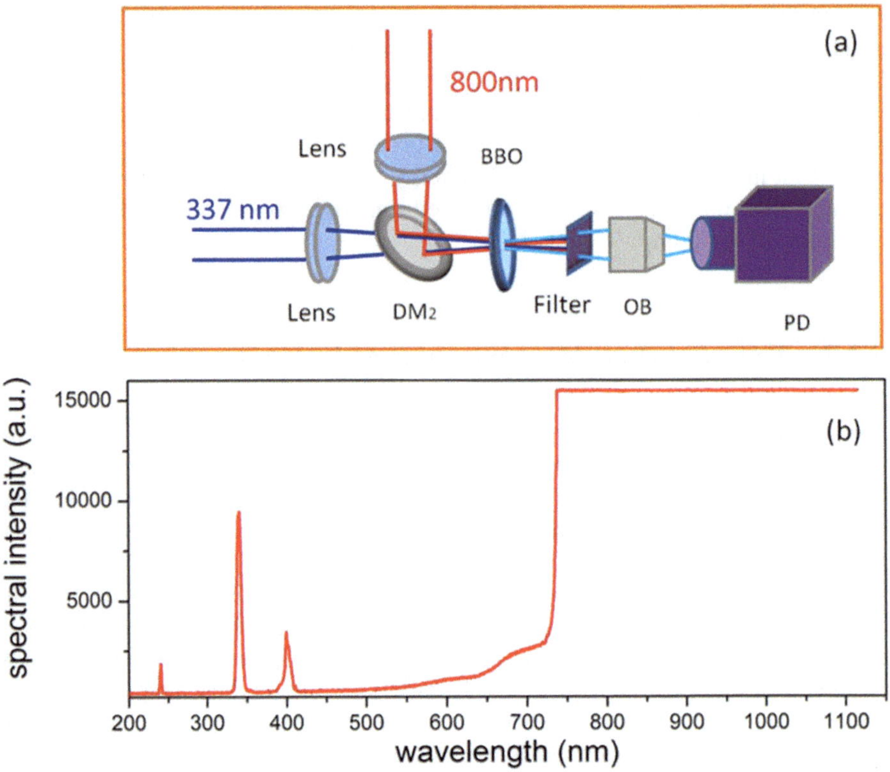

Fig. 3.8 (**a**) Schematic of the setup for the cross correlation measurement. (**b**) Spectrum of the SFG process

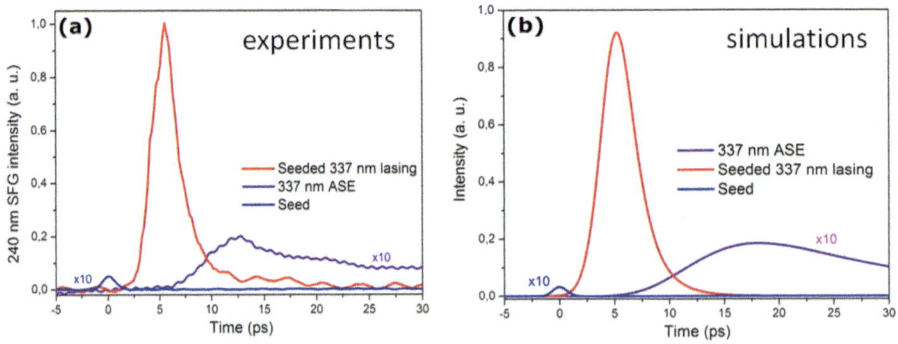

Fig. 3.9 Temporal profiles of the forward-propagating seed pulse (*blue*), externally seeded 337 nm lasing pulse (*red*) and 337 nm ASE (*violet*). (**a**) Measurements using the cross correlation method, (**b**) simulations

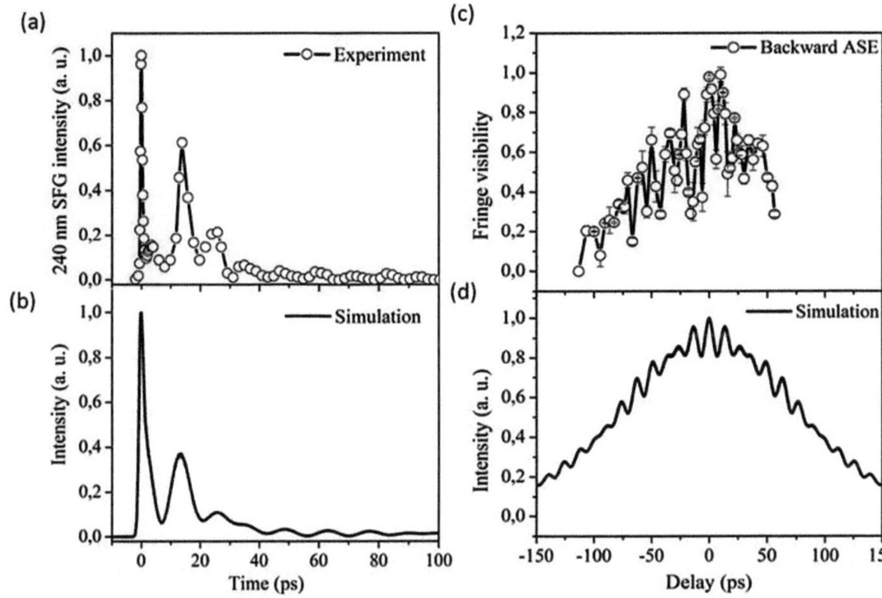

Fig. 3.10 (**a**) Time-resolved measurement of the backward-seeded 337 nm emission. (**b**) Results of a corresponding numerical simulation. (**c**) Autocorrelation measurement of the backward-propagating ASE. (**d**) Corresponding numerical simulation

respect to the seed pulse, and a pronounced intensity modulation with the period of 11 ps is evident. The strongest pulse that is following the seed pulse has a width (FWHM) of 4 ps. By measuring the lasing pulse energy with a photodiode, we found that the backward-seeded emission was about 150 times weaker than the forward emission under the same excitation conditions.

Since backward-propagating unseeded 337 nm ASE is much weaker than the ASE in the forward direction or the backward-seeded lasing, the above cross correlation method cannot be employed for its characterization. Instead, we used a Michelson interferometer-based setup, shown in Fig. 3.11, to evaluate the degree of coherence of the backward-propagating ASE. The interference pattern of the 337 nm lasing beam was captured by an iCCD camera (Princeton Instruments, PI-MAX). The fringe contrast is defined as $(I_{max} - I_{min})/(I_{max} + I_{min})$, where I_{max} and I_{min} are the maximum and minimum intensities, respectively. In the experiments, the temporal delay between the two replicas of the signal was changed, and the fringe contrast was measured, which provided a field autocorrelation measurement of the 337 nm radiation.

Figure 3.10c shows the interference fringe patterns of the backward 337 nm ASE for different interferometer delays. The fringe visibility persists for the delay of about 80 ps. Therefore, the pulse duration of the backward 337 nm ASE can be estimated as 80 ps/1.5 ≈ 50 ps. We note that in this case, a periodic oscillation of the fringe structure with a period of ~11 ps is visible.

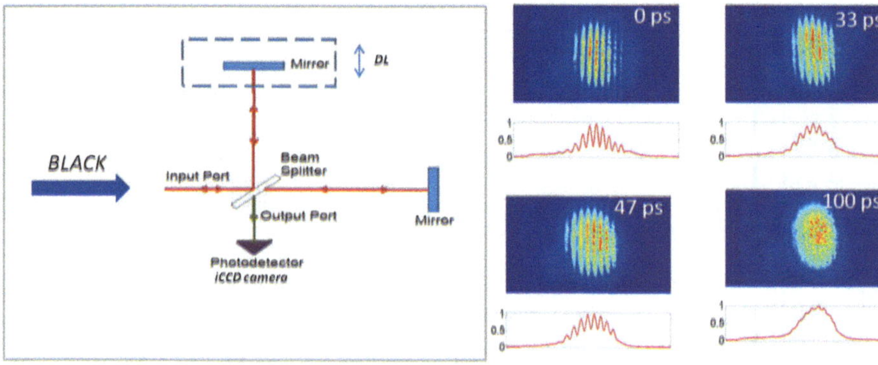

Fig. 3.11 *Left*: Field autocorrelation measurement of the backward-unseeded 337 nm ASE using a Michelson interferometer. *Right*: Interference patterns recorded by an iCCD camera for different interferometer delays indicated in the patterns

What is the origin of this temporal modulation of the lasing observed mainly for the backward-unseeded emission? The lasing line at 337.1 nm, which corresponds to the P branch of the $C^3\Pi_u^+ \rightarrow B^3\Pi_g^+$ transition, is composed of three fine-structure branches P_1, P_2, and P_3, which are due to the interaction between the orbital angular momentum and spin [33, 34]. The separation between these lines is about 0.33 Å [33], which is beyond the resolution of our spectrometer. The fast collisional excitation process brings the molecules into a coherent superposition of the three branches. An intensity modulation in the temporal domain is then expected due to the quantum beating between them. The period T of the corresponding temporal modulation can be expressed as $T = \lambda_0^2 / (c \cdot \Delta\lambda)$, where $\lambda_0 = 337.1$ nm is the central wavelength of the lasing emission, c is the light speed, and $\Delta\lambda = 0.033$ nm is the wavelength separation of the three branches. Therefore, one obtains $T = 11.5$ ps, in good agreement with the experiment result.

3.2.4 Modeling and Numerical Simulations of N_2 Lasing

To understand the temporal dynamics of the lasing discussed above, numerical simulations using non-adiabatic Maxwell-Bloch equations have been conducted, following earlier studies [35, 36]. The one-dimensional, time-dependent Maxwell-Bloch code DeepOne [37], originally developed to study the amplification of soft X-rays (10–40 nm) in hot, dense plasmas ($n_e > 10^{18}/\text{cm}^3$; $T_e > 10$–100 eV), has been adapted to the problem of nitrogen lasing. The code solves the paraxial wave equation for electric field in the slowly varying envelope approximation:

$$\frac{\partial E_\pm}{\partial t} \pm c \cdot \frac{\partial E_\pm}{\partial z} = \frac{i\omega}{2}\left[\mu_0 c^2 P_\pm - \left(\frac{\omega_p}{\omega_0}\right)^2 E_\pm,\right] \tag{3.4}$$

where E_+ and E_- are the electric field envelopes propagating in the forward and backward directions, respectively, P_+ and P_- are the corresponding polarization densities of the medium, c is the speed of light, ω_0 is the frequency of electric field, ω_p is the plasma frequency, and μ_0 is the permeability of vacuum. Equations (3.4) are coupled to the following equations for polarizations that are derived from Bloch equations:

$$\frac{\partial P_\pm}{\partial t} = \Gamma - \gamma P_\pm - \frac{iz_{ul}^2}{h} E_\pm \left(N_u - N_l\right), \tag{3.5}$$

where Γ is a stochastic source term with vanishing correlation time that models spontaneous emission [38], γ is the depolarization rate due to collisions, z_{ul} is the corresponding dipole matrix element, and N_u and N_l are the populations of the upper and lower levels of the lasing transition, respectively. These populations are computed by using rate equations

$$\frac{\partial N_i}{\partial t} = \sum_k C_{ki} N_k \pm Im\left(E \times P\right)/2h, \tag{3.6}$$

where the index i marks the upper (u) and lower (l) states, and C_{ki} are the collisional excitation and radiative de-excitation rates. These rates are computed from the cross sections reported in [39]. The evolution of the electron density and temperature is computed using the model presented in [19, 35, 36].

Figure 3.9b presents the simulated temporal profiles of the forward ASE and of the seeded emission. The agreement between simulation and experimental results is quite remarkable. The seed pulse is amplified 188 times, and the maximum of the amplified pulse is delayed by 5.35 ps. The FWHM duration of the amplified pulse is 3.6 ps. Simulation and experiments also agree on the intensity and pulse duration of the forward 337 nm ASE. It is found to be 45 times less intense than the seeded emission that has a FWHM duration of 20.79 ps.

The temporal evolution of the lasing pulse along the propagation direction was also computed. The results are shown in Fig. 3.12. The 337 nm ASE starts from noise spanning the entire gain lifetime corresponding to the lifetime of the upper level of the lasing transition $\tau_u \sim 600$ ps. Amplification and saturation effects shorten the duration of the pulse. After passing through 3 cm of plasma, the ASE pulse has a FWHM duration of 20.8 ps. The evolution of the temporal structure of the amplified external seed pulse in the filament plasma amplifier, shown in Fig. 3.12b, is mainly driven by the initial spectral profile of the seed pulse and the linewidth of the amplifier, dominated by collisions (i.e., a Lorentzian line). When the seed pulse is intense enough, saturation effects reduce the duration of the pulse until it attains its value of 3.6 ps (FWHM) after propagating through 3 cm of plasma.

Figure 3.10b shows the simulated temporal evolution of the seeded backward emission. Note that the seed and the IR pump pulses enter the plasma from opposite sides, at $t = 0$ ps. Thus, the seed propagates without amplifying until it meets the IR pump in the middle of the filament ($t = 50$ ps). From this point on, the seed is amplified.

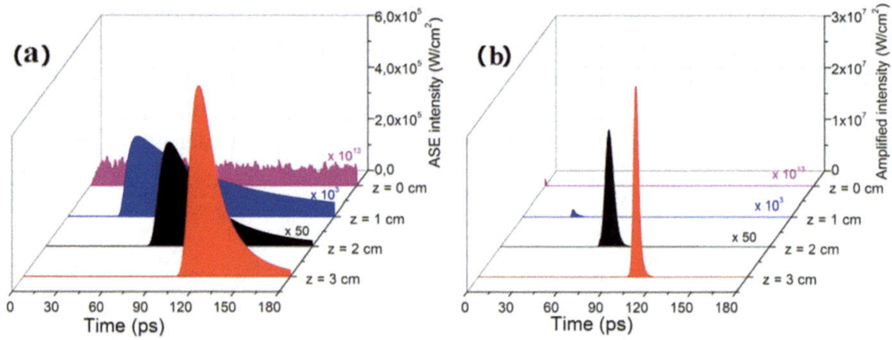

Fig. 3.12 Evolutions of temporal profiles of the ASE lasing pulse (**a**) and external seed pulse (**b**) during amplification along the filament plasma in the forward direction. For better comparison, some curves are scaled by different factors, which are specified next to the curves

At the exit from the filament amplifier, the simulation predicts a moderately amplified seed pulse, followed by a wake radiation extending over more than 60 ps, in good agreement with our experiment observations. We point out that this kind of wake emission has been widely observed in the simulation of X-ray amplification in gas plasma [40].

Backward 337.1 nm ASE is much weaker than the forward one. This is due to the particular geometry of the longitudinal pumping. When spontaneously emitted photons co-propagate with the IR pump beam, they interact continuously with newly created population inversion. On the contrary, in the case of backward ASE, where the signal counter-propagates with respect to the IR pump, population inversion experienced by the spontaneously emitted photons is gradually depleted by the forward ASE that follows the IR pump pulse. In fact, due to the short lifetime of the gain τ_g = 13 ps, the effective amplification length for the spontaneously emitted photon propagating in the backward direction is $l = c\tau_g$ = 4.2 mm, which is much less than the 30 mm geometrical length of the plasma filament [15]. After propagating through 30 mm of the N_2 plasma, the pulse duration (FWHM) of the backward ASE is found to be 77 ps, in good agreement with the autocorrelation measurements shown in Fig. 3.10c.

3.2.5 *Conclusions*

To conclude this part of the chapter that discussed air lasing in neutral nitrogen molecules, we have demonstrated that stimulated emissions at 337 nm in both backward and forward directions can be obtained from plasma filaments pumped by circularly polarized 800 nm femtosecond pulses. An external seed pulse injected into the gain channel in either forward or backward direction can be amplified by two orders of magnitude. Using cross correlation and autocorrelation techniques, we have characterized these emissions in time domain. Important differences between intensity and duration of backward and forward emissions are observed, for both amplified spontaneous emission (ASE) and seeded amplification. Numerical simulations based on the non-adiabatic Maxwell-Bloch equations reproduce our observations and explain these differences by the finite gain lifetime and by the traveling mode of excitation in this gas laser.

3.3 Lasing in Singly Ionized Nitrogen Molecules: Superradiance and the Role of Electron Recollisions

In 2011, an intense forward-propagating radiation at 391 nm and other transitions in singly ionized nitrogen molecular ion produced through photoionization by intense mid-infrared laser pulses have been reported by the group at the Shanghai Institute of Optics and Fine Mechanics [2]. The effect corresponded to the $B^2\Sigma_u^+ \rightarrow X^2\Sigma_g^+$ transition between the excited and ground states of the nitrogen cation. The pump source was a femtosecond tunable optical parametric amplifier (OPA) operating in the mid-infrared (1.2 – 2.9 μm). Whenever third or fifth harmonic of the driver pulse overlapped with a particular transition in N_2^+, strong narrow-linewidth radiation, superimposed on the relatively broad spectrum of the harmonic, was observed, as shown in Fig. 3.13.

Soon after, it has been reported that such a narrow-linewidth emission, which corresponded to the $B^2\Sigma_u^+ \rightarrow X^2\Sigma_g^+$ transition in N_2^+, can also be observed with pump pulses at various wavelengths including 800 nm [15, 41], 400 nm [20], 1.03 and 3.9 μm [22], and 1500 nm [23]. The experimental configuration has been extended from the one involving a single pump pulse to the seeded configuration [15] and to pumping by an adaptively controlled sequence of pulses [22].

Lasing in N_2^+ has several unusual features such as an ultrafast gain buildup, strong dependence on the polarization of the pump light, and superradiance. Some

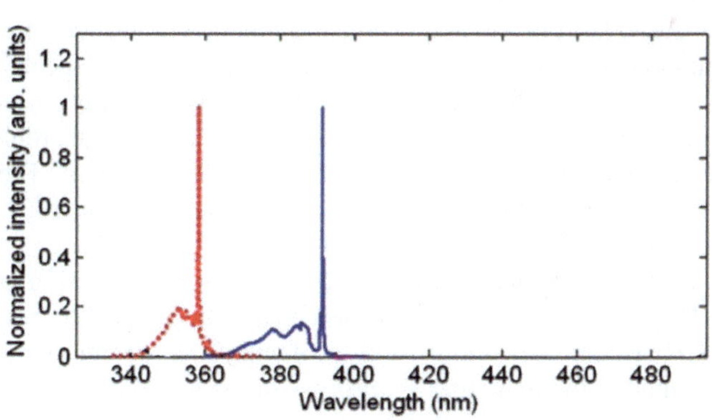

Fig. 3.13 Spectra of laser-like emissions at 391 and 357 nm generated in air plasma when pumped at 1920 and 1760 nm, respectively (Reproduced from [2] with permission from authors)

of those features will be discussed in more detail below. Until now, the physical mechanism responsible for the generation of optical gain in this system remains to be controversial and is actively discussed in the filamentation and strong-field communities. In the original report [2], the effect was explained by amplification of the weak harmonic seed pulse in the gain medium with population inversion. It was later found that the effect depends on the intensity of the pump pulse and on the pressure of the gas. Under some conditions, gain for an external seed pulse is only observed with pumping at 800 nm, but not with mid-IR pumping [6].

In what follows, we will briefly introduce the main features of the N_2^+ lasing when it is pumped by 800 nm pulses. We will then discuss various proposed physical mechanisms that have been suggested to explain the presence of gain. Time-resolved measurements of the emission as a function of the gas pressure will be discussed next. We will argue that our results point to the superradiant emission. We will further present experimental results on the N_2^+ lasing pumped by an 800 nm laser pulses with different polarization ellipticity and discuss the possible role of electron recollisions in this lasing effect. Experimental results on the fine dependence of the lasing on the center wavelength of the pump source support the essential role of electron recollisions as the dominant mechanism responsible for the creation of population inversion in N_2^+.

3.3.1 Main Features of the Nitrogen Ion Lasing

3.3.1.1 Self-Seeded Lasing in the Case of 800 nm Femtosecond Laser Pumping

In the first report on the use of 800 nm pumping for the initiation of lasing in N_2^+ [41], an external seed pulse at either 391 or 428 nm was required for the lasing to be observed. It has been later found that under certain conditions the seed pulse was not necessary. In Fig. 3.14, we show spectra of the lasing in pure nitrogen, when pumped by focused laser pulses at 800 nm, as a function of nitrogen pressure. At the lowest pressure $p = 4$ mbar, a weak and spectrally broad emission peaking at 400 nm is observed, as shown in the inset. As the pressure is increased, lasing radiation around 391 nm appears and grows with pressure to a maximum at around 100 mbar. For pressure values above 200 mbar, the 391 nm lasing disappears. The intensity of the lasing line around 428 nm increases significantly from 100 mbar up to $p = 750$ mbar and then saturates for $p > 750$ mbar [15].

We interpret the weak and spectrally broad radiation around 400 nm, shown in the inset in Fig. 3.14, as being due to the generation of second harmonic of the pump. Second harmonic generation in laser filaments has been reported in the past [42] and attributed to the nonlinear $\chi^{(3)}$ process in the presence of the static electric field due to ionization. We suggest that this second harmonic serves as the seeding pulse to start the lasing around 391 nm in the low-pressure regime between 8 and 200 mbar. White-light continuum becomes visible to the human eye at pressure

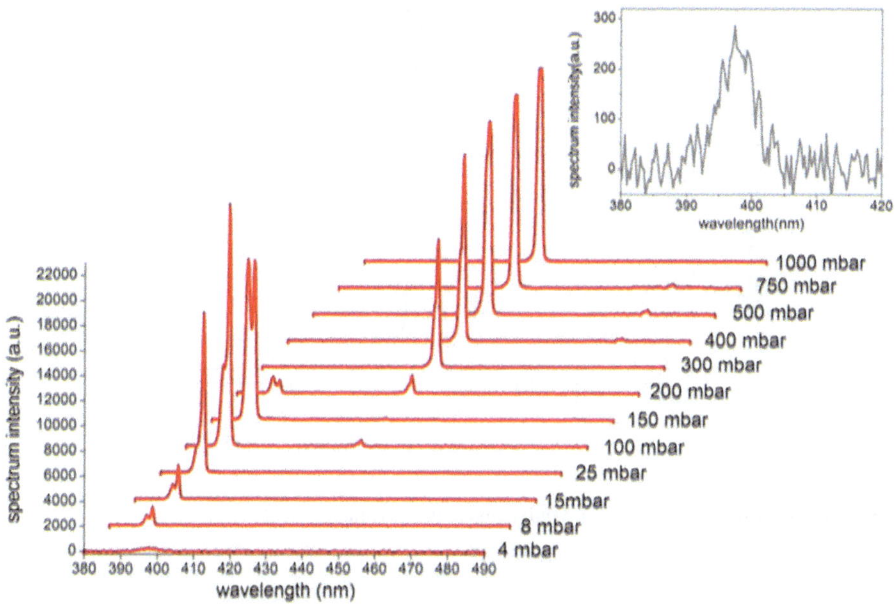

Fig. 3.14 Spectra of the forward-propagating radiation generated in pure nitrogen, as a function of nitrogen pressure. The inset shows a zoomed spectrum around 400 nm at 4 mbar of pressure

exceeding 100 mbar. The continuum does not appear in Fig. 3.14 because it is strongly attenuated by color glass filters in the detection system. We therefore believe that the supercontinuum generated in the relatively high-pressure regime provides the seeding pulse for the 428 nm emission. Similar results have been reported in [20, 43]. The discovery of the self-seeded regime of lasing in N_2^+ under 800 nm pumping made the studies of this effect easily accessible to the ultrafast optics community. However, as we will discuss below in Sect. 3.3.4, the exact value of the center wavelength of the pump, around 800 nm, is an important parameter that may affect the lasing significantly.

3.3.1.2 Amplification of a Seed Pulse in the N_2^+ Plasma

A seed pulse at the appropriate wavelength, when injected into the plasma, can be significantly amplified, as shown in Fig. 3.15. Here, the femtosecond pump pulse is focused by a convex lens with the focal length $f = 400$ mm in pure nitrogen at the pressure of 30 mbar. The seed pulse around 391 nm is generated by second harmonic generation in a thin BBO crystal. The weak seed pulse is combined with the 800 nm pump pulse via a dichromatic mirror. The seed pulse is also focused by an $f = 400$ mm lens installed before the dichroic mirror. The pump pulse energy is 3 mJ. The self-seeded stimulated emission is responsible for the weaker signal observed when no seed pulse is applied.

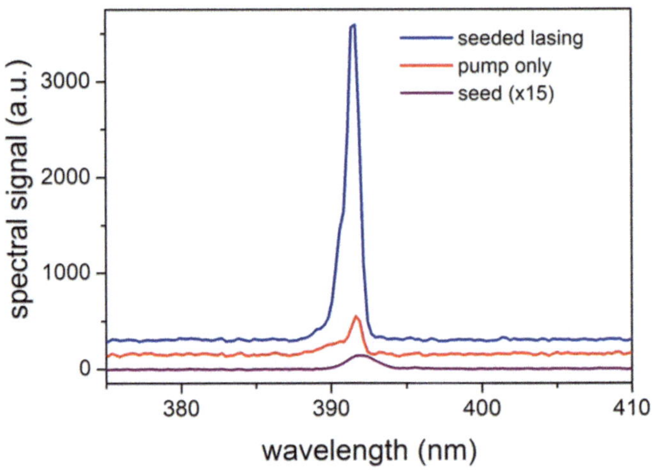

Fig. 3.15 Amplification of the seed pulse inside the plasma filament

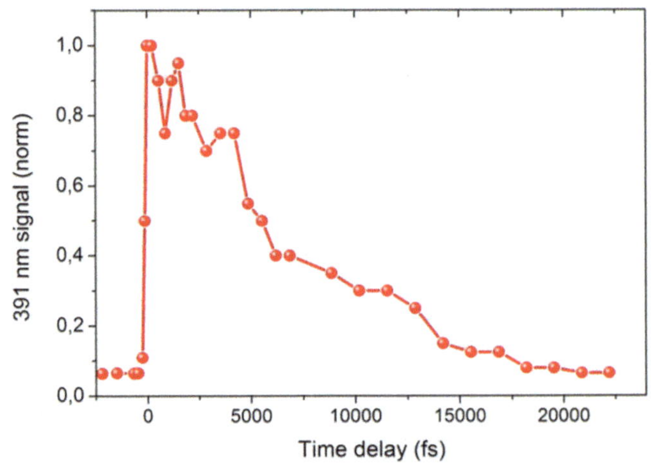

Fig. 3.16 Temporal evolution of the optical gain at 391 nm measured in 10 mbar pure nitrogen gas as a function of the delay between pump and seed pulses

3.3.1.3 Ultrafast Gain Buildup

One of the puzzling aspects of the lasing action in N_2^+ is the ultrafast gain buildup time. In [5], the dynamics of the optical gain is measured by recording the amplified 391 and 428 nm signals as a function of the time delay between the pump and seed pulses. The result obtained in 10 mbar of pressure of pure nitrogen gas is shown in Fig. 3.16. A very rapid (<100 fs) raising edge of the gain buildup is evident. This ultrafast gain buildup excludes the electron impact excitation as a possible gain mechanism, because electron-molecule collision time is on the order of ~10 ps at this gas pressure. This result further indicates that the optical gain is produced within the duration of the laser pulse (45 fs). Therefore, either a direct strong-field process or a photon-assisted electron process, such as electron recollision, must be responsible for the formation of the optical gain.

3.3.2 *Controversy About the Mechanism of the* N_2^+ *Lasing*

The physical mechanism underlying the lasing effect in N_2^+ is currently the subject of a controversy. In the first report of this effect [2], where mid-IR laser was employed to pump lasing in ambient air, the authors suggested that the emissions were due to the amplification of the third or fifth harmonic of the pump beam in the medium with population inversion. It was argued that the pump harmonics, in addition to supplying the seed pulse, were instrumental in emptying the lower emission state. Soon after, forward-propagating emissions at 391 and 428 nm were observed with pumping at 800 nm [15, 20, 41]. Very recently it was argued that 800 nm and mid-IR pumping regimes are fundamentally different. While significant amplification of an external seed pulse under 800 nm pumping has been reported by several groups, no amplification of an external seed was so far observed with a mid-IR pumping [6]. (With mid-IR pumping, so far the emissions were only observed in the self-seeded regime [2, 23].) Based on these observations, very recently it was argued that the lasing in N_2^+ is due to the resonantly enhanced Raman amplification [44]. In what follows, we will briefly discuss various alternative suggestions for the gain mechanism responsible for lasing in N_2^+.

3.3.2.1 "Natural" Population Inversion

It has been suggested that, in the presence of the fundamental and the self-generated harmonic photons, population inversion can be established in any small molecular ion possessing ionic states that are energetically close to the ionic ground state [21]. In the case of N_2^+, it was proposed that a three-photon absorption from the ground state $X^2\Sigma_g^+$ to the excited state $B^2\Sigma_u^+$, or a four-photon process involving higher vibrational levels of the excited state, may give rise to gain. It was suggested that this mechanism of population inversion is universal and the population inversion thus created was termed "natural population inversion." The experimental confirmation of this proposal is presently lacking.

3.3.2.2 Population Inversion Enabled by the Post-ionization State Coupling

Very recently, two groups have performed numerical simulations of the quantum dynamics of the N_2^+ ion in the strong laser field [4, 7]. Both groups argued that the intermediate state $A^2\Pi_u$ plays the key role in establishing population inversion between the $B^2\Sigma_u^+$ and $X^2\Sigma_g^+$ states. Immediately following ionization, the N_2^+ ion is predominantly initiated in the ground ionic state $X^2\Sigma_g^+$. As the simulations based on the time-dependent Schrödinger equation, reported in [4, 7], show, the subsequent population dynamics, which is due to the state coupling by the strong laser field, redistributes the ionic population via Rabi oscillations. For sufficiently high laser intensity, the calculations show that by the end of the pump pulse, most of the population ends up in the intermediate state $A^2\Pi_u$, resulting in the net population inversion between the $B^2\Sigma_u^+$ and $X^2\Sigma_g^+$ states. The intermediate state $A^2\Pi_u$ acts as a population reservoir. However, the simulations also show the net population inversion between the $A^2\Pi_u$ and $X^2\Sigma_g^+$ states, but no stimulated emission on the $A^2\Pi_u \rightarrow X^2\Sigma_g^+$ transition has been so far observed. (The corresponding emission wavelength would be 1.1 μm.)

3.3.2.3 Inversion-Free Amplification due to the Transient Molecule Alignment

It was suggested that the impulsive alignment of the nitrogen molecular ions plays an important role in the N_2^+ lasing [23, 45, 46]. Specifically, pump-probe measurements of the gain dynamics revealed strong variations of gain for the pump-probe delays near the instances of impulsive rotational revivals [45]. In the experiments on nitrogen lasing pumped by adaptively controlled pulse sequences at 1030 nm wavelength, a significant enhancement of the N_2^+ emissions at 391, 258, and 428 nm was observed when the delay between the individual pulses in the sequence matched the rotational revival time [46].

Based on these observations, it was suggested that a transient gain may occur without actual population inversion between the upper and lower energy levels of the lasing transition in N_2^+. The probability of emission is proportional to the population of the upper state times the alignment factor for the molecular ions in the upper state $\cos^2(\theta)_B$, while the probability of absorption is proportional to the population in the lower state times the alignment factor for the ions in the lower state $\cos^2(\theta)_X$. Gain results when the probability of emission is higher than the probability of absorption. Due to the slight difference in the fundamental rotational periods of the N_2^+ molecular ions in the lower energy state $X^2\Sigma_g^+$ and the higher energy state $B^2\Sigma_u^+$, the alignment factor for the ions in the upper state can be momentarily large at the same time when the alignment factor for the ions in the lower state is small. Then, even though the population inversion between the upper and lower levels is negative (no inversion), gain may exist during those transient gain windows.

The above mechanism predicts gain during a series of short temporal windows around the instances of molecular alignment revivals [46]. It explains the strong variation of gain in pump-probe experiments, when the pump-probe delay is close to the instances of rotational revivals, which has been observed by several groups. However, this mechanism cannot explain the sustained gain that persists, on top of the transient gain windows, for up to ten picoseconds, which also has been widely observed. We believe that the transient molecular alignment certainly plays an important role in the N_2^+ lasing, but it is not the essential effect responsible for gain in this system.

3.3.2.4 Rotational Population Inversion Mechanism

Very recently, another gain mechanism has been proposed that is also based on rotations of nitrogen molecular ions [23]. Unlike the coherent mechanism discussed above, which is active only within a sequence of short temporal windows, this gain mechanism is incoherent and can explain the sustained gain that persists for several picoseconds. This gain mechanism relies on different polarizabilities of the nitrogen molecular ions in their upper energy state $B^2\Sigma_u^+$ and lower energy state $X^2\Sigma_g^+$. As a result, the ions in the two states are rotated differently by the pump field, resulting in different rotational distributions in the upper and lower electronic states. Even though the net population inversion between the upper and lower electronic states can be negative (no inversion), inversion may exist between the subset of the rotational transitions within the electronic transition manifold $B^2\Sigma_u^+ \rightarrow X^2\Sigma_g^+$. This subset of rotational transitions generates gain.

The above gain mechanism has been demonstrated in low-pressure pure nitrogen gas pumped by mid-IR femtosecond laser pulses at 1.5 μm wavelength. By high-resolution spectroscopy of both stimulated and spontaneous emissions, it has been shown that the upper and lower electronic states in N_2^+ have distinctly different rotational population distributions. Separate measurements of stimulated and spontaneous emission spectra were essential in this study, as the measurement of stimulated emission only, which has been routinely done previously, bears information about the difference in rotational population distributions between the upper and lower emission states. Based on the known spin statistics of nitrogen, the peak value of the total gain in the 1-mm-long interaction zone was estimated as 64, corresponding to the lower-bound estimate for the peak gain per unit length of about 40 cm^{-1}. Since the relaxation of rotational distributions to the thermal distribution involves collisions of N_2^+ ions with electrons, this gain mechanism has picosecond-scale lifetime.

3.3.2.5 Superradiant Behavior

The first report of the signatures of superradiance in the N_2^+ lasing has been by the SIOM group in 2014, based on their time-resolved measurements of the emission as a function of the plasma length and gas pressure [47]. Systematic experiments that followed have confirmed the superradiant nature of the 391 nm emission. In those experiments, the temporal profile of the forward-propagating 391 nm radiation was measured by sum-frequency generation (SFG) of the 391 nm signal and a reference 800 nm pulse in a BBO crystal. The SFG signal at 263 nm was recorded as a function of the relative delay between the signal and reference pulses. The results are

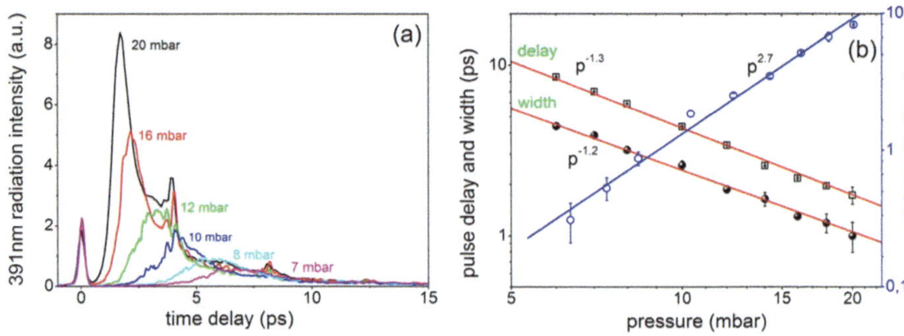

Fig. 3.17 (**a**) Temporal profile of the forward-propagating 391 nm radiation measured at different nitrogen pressures in the presence of a constant seed pulse. The seed pulse shows as the narrow peak at zero delay. (**b**) The pulse delay, pulse width, and the 391 nm peak intensity as a function of nitrogen pressure.

shown in Fig. 3.17a. The peak power of the radiation increases significantly with gas pressure, accompanied by the simultaneous reduction of both the pulse width and pulse delay. All of those features are characteristics of superradiance.

It is well known that the delay and duration of a superradiant pulse both scale as $\propto N^{-1}$, while the power of the pulse scales as $\propto N^2$, where N is the number of emitters, which is directly proportional to the gas pressure [48]. The data for the delay, pulse duration, and the peak power of the emission, together with best fits, are shown in Fig. 3.17b. The overall agreement between the data and the theoretical expectation for superradiance confirms the superradiant nature of the 391 nm lasing.

3.3.3 A Clue Supporting Electron Recollision Excitation: Strong Dependence on the Polarization Ellipticity of the Pump

Recent experiments on the dependence of the N_2^+ lasing on the polarization ellipticity of the pump beam provided an important clue to the physical mechanism underlying the emission. The experimental results for the 391 and 428 nm emissions are shown in Fig. 3.18a. The signal is strongest for the linearly polarized pulse, and it is suppressed as soon as the degree of ellipticity ε exceeds 0.3. Similar dependence has been previously reported by the SIOM group [49]. Here, we notice that this dependence bears a striking similarity to the behavior of high harmonic generation (HHG) [50]. Note that there is a small dip around $\varepsilon = 0$ for the 391 nm signal. A similar dip has been reported for the HHG with photon energies that are close to the ionization potential in oxygen, hydrogen, and noble gases [51–53]. We have also measured the HHG yield with photon energy close to the molecular ionization

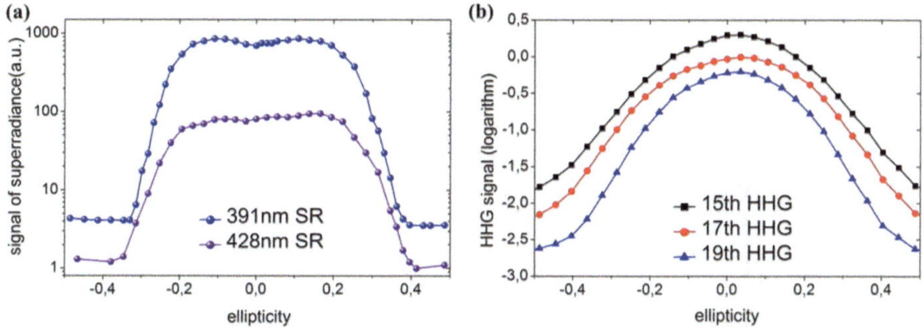

Fig. 3.18 (**a**) 391 and 428 nm lasing emissions as a function of polarization ellipticity of the pump pulses at 800 nm. (**b**) Dependence of the high-order harmonic yield in nitrogen gas as a function of the polarization ellipticity of the laser. The experiments were performed in a 15-mm-long gas cell with two 150 μm holes in the entrance and exit cell facets. The holes were drilled in the 100-μm-thick aluminum foil windows by the laser beam itself

potential under identical experimental conditions. The results of that measurement are shown in Fig. 3.18b.

The effect of the polarization ellipticity of the pump in high harmonic generation is well understood. A semiclassical model predicts the main features remarkably well [50]. The HHG process is divided into three successive steps: tunnel ionization, motion of the electron wave packet in the strong laser field, and the recombination with the parent ion. With a circularly polarized driver pulse, the returning free-electron wave function never overlaps with the parent ion and therefore cannot emit its kinetic energy in the form of high harmonics.

Applying the same semiclassical model to our case, we interpret the ellipticity dependence of the gain as being due to the non-radiative transfer of ion population from $X^2\Sigma_g^+$ to $B^2\Sigma_u^+$ state via laser-field-assisted recollision. In each pumping event, the electron in the presence of the intense pump field is removed from the outer orbital of the neutral nitrogen molecule. Then it is accelerated and driven back to the parent molecular ion by the laser field. There it inelastically collides with an electron occupying an inner orbital of the N_2^+ ion. In the recollision, if the impacting electron has a sufficient energy, there is a certain probability to transfer the inner-orbital electron to the outer orbital of the molecular ion. The result of this reaction is an excited molecular ion and a free electron.

3.3.4 *Confirming the Essential Role of Recollisional Excitation*

In the recollisional excitation process, the neutral nitrogen molecule has a certain probability to be ionized within each optical cycle. The electron is released from the molecule with zero initial velocity. In the presence of intense electric field of the

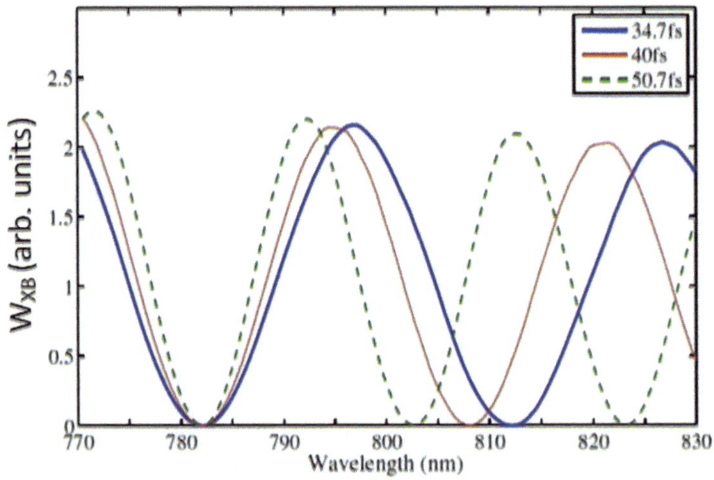

Fig. 3.19 Calculated probability of collisional excitation of the $B^2\Sigma_u^+$ state as a function of the wavelength of the pump laser

laser, the electron is alternately accelerated and decelerated around its parent ion. Once the electron inelastically collides with the parent ion, which is assumed to be in its ground state, the collision can lead to the excitation of the ion to its excited state (the $B^2\Sigma_u^+$ state in the case of N_2^+ lasing). The probability that the ion is left in the $B^2\Sigma_u^+$ state can therefore be obtained by integrating those inelastic recollision events over the pulse duration, taking into account the evolution of the phase of the excitation probability. The latter reads as follows:

$$W_{XB} = V_0^2 \left| \sum_{\tau_0} sign\left(v_e^s\right) w\left(\tau_0\right) \exp\left[-2\pi i \tau_s \left({}^{\omega_{XB}}\!/_{\omega_0} \right) \right] \right|^2, \tag{3.7}$$

where $w(\tau_0)$ is the ionization probability at time τ_0, $\tau_s(\tau_0)$Z is the first recollision time for the electron born at time τ_0 normalized to the optical cycle, and sign$\left(v_e^s\right)$ is the sign symbol determined by the direction of motion of the electron at the time of the recollision, equal to +1 or −1. Note that a one-dimensional recollision model is assumed here for simplicity.

Expression (3.7) has been evaluated numerically assuming pump pulse with a constant amplitude. The velocities and trajectories of the electrons were obtained from Newton's equations of motion. Only those electrons that have acquired sufficient kinetic energy $E > 3.17$ eV in the laser field to enable a collisional transition $X^2\Sigma_g^+ \rightarrow B^2\Sigma_u^+$ were considered. The predicted dependence of the transition probability on the wavelength of the pump field is shown in Fig. 3.19 for three durations of the pump pulse. The calculation reveals three features of this transition probability. First, it exhibits cyclic variations vs. wavelength. Second, the period of these

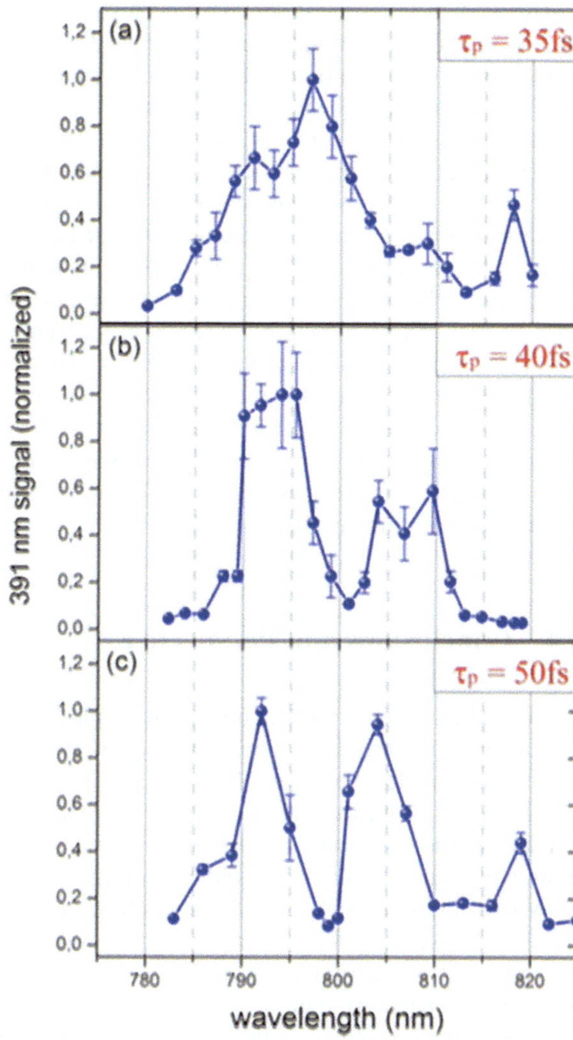

Fig. 3.20 Measured forward-propagating signal at 391 nm as a function of wavelength of the pump laser for three different pulse durations, as shown in the individual figures

undulations depends on the duration of the pump pulse. The period is shorter if the pump pulse is longer. Third, the model predicts zero probability of excitation for any pulse duration when the laser wavelength is tuned to 782 nm, where its second harmonic frequency matches the frequency of the 391 nm transition. The latter is due to the opposite phases (0 or π) that the recolliding electrons have when they approach the parent ion from opposite directions. In another words, the probabilities of excitation by the recollisions with the electrons coming from the opposite sides of the parent ion interfere destructively within each optical cycle, when the resonance condition $\omega_0 = \omega_{BX}/2$ is satisfied. We would like to point out that this probability cancelation is similar to the fact that no second (or any other even-order) harmonic can be generated in an isotropic medium.

The predictions of the model have been verified experimentally in collaboration with N. Ibrakovic, S. Bengtsson, A. Cord, and A. L' Huillier from Lund University in Sweden, using a femtosecond laser source whose center wavelength can be tuned from 780 to 820 nm. The laser beam was focused into a gas cell with pure nitrogen at 5 mbar pressure. The numerical aperture of the beam was 0.025. The forward-emitted signal at 391 nm was detected by a photodiode behind a narrow-band color filter passing the 391 nm emission. The experimental results are shown in Fig. 3.20. All three main features of the excitation predicted by the model were confirmed. Specifically, the signal had a dip at the resonant wavelength around 782 nm, exhibited cyclic variations vs. the excitation wavelength, and the period of those undulations decreased with increasing duration of the pump pulse. These results together with the dependence of the lasing on the polarization of the pump pulse confirm that recollisions are essential for the N_2^+ lasing. Recent additional measurements and calculations detailed in the October 2017 issue of Physical Review Letters show that it is the macroscopic dipole induced by recollisions that is responsible for the lasing, noyt the population transfer. Also, the 1-d recollision model used here has been generalized to 3 dimensions (see V. Tikhochuk et al EPJD 2017).

3.4 Conclusions and Perspective

Stimulated emission from N_2 and N_2^+ in air plasma came as a surprise for the filamentation and strong-field communities. For the ASE in N_2, the recently demonstrated scheme for backward lasing holds promise for future applications. Its underlying mechanism, which has been confirmed both experimentally and theoretically, is the collisional excitation of the nitrogen molecules by energetic electrons, which can be obtained by either circularly polarized near-infrared laser pulses or by mid-infrared laser pulses of any polarization. We have performed numerical simulations to reveal the temporal dynamics of this lasing effect and obtained good agreement with experiments. The significant difference between the forward- and backward-propagating emissions has been shown to originate from the traveling excitation scheme and from the limited lifetime of optical gain. In the future, the quenching effect of oxygen molecules on the lasing in neutral nitrogen needs to be studied. This quenching is currently one of the major roadblocks toward the demonstration of backward nitrogen lasing in real air.

The forward-propagating lasing in N_2^+ still remains mysterious 6 years after its discovery. This effect has several unusual features such as an ultrafast gain buildup on the femtosecond time scale, strong dependence on the ellipticity of the pump light, and superradiance. Different mechanisms to explain gain in this system have been proposed such as population inversion due to post-ionization state coupling, laser without inversion enabled by transient molecular alignment, and gain through rotational excitation of nitrogen molecular ions. Based on the strong dependence of the lasing yield on the ellipticity of the pump laser, we speculate that the electron recollisional excitation plays an important role in the establishment of gain.

Numerical simulations based on the recollisional excitation predict cyclic variations of the lasing signal with the wavelength of the pump light. Recent experiments with a wavelength-tunable femtosecond laser system confirm that prediction. This leads us to conclude that recollisions indeed play an essential role in the lasing. However, this problem is far from being conclusively resolved. In the view of the rich physics involved and the disparate proposals for the essential gain mechanisms, this phenomenon will attract a growing attention in the coming years.

Acknowledgments We would like to acknowledge fruitful collaborations with Eduardo Oliva of Madrid Technical University (Spain), Shihua Chen of Southeast University (China), Nevel Ibrakoivc, Samuel Bengtsso, Cord Arnold, Hohan Mauritsson, and Anne L'Huiller of Lund University (Sweden), Arnaud Couairon of Ecole Polytechnique (France), Rostyslav Danylo of Ecole Polytechnique (France), Sergey Mitryukovskiy of Russian Quantum Center (Russia), and Vladimir Tikhonchuk of Bordeaux University (France). Y. Liu would like to thank Hongbing Jiang and Chengyin Wu of Peking University (China), Ya Cheng and Jinping Yao of SIOM (China), and Huailiang Xu of Jilin University (China) for the stimulating and fruitful discussions.

References

1. A. Dogariu, J.B. Michael, M.O. Scully, R.B. Miles, Science **331**(6016), 442–445 (2011)
2. J. Yao et al., Phys. Rev. A **84**, 051802(R) (2011)
3. S. Mitryukovskiy, Y. Liu, P. Ding, A. Houard, A. Mysyrowicz, Opt. Express **22**(11), 12750–12759 (2014)
4. H. Xu, E. Lotstedt, A. Iwasaki, K. Yamanouchi, Nat. Commun. **6**, 8347 (2015)
5. Y. Liu et al., Phys. Rev. Lett. **115**, 133203 (2015)
6. J. Yao et al., Phys. Rev. Lett. **116**, 143007 (2016)
7. A.J. Traverso et al., Proc. Natl. Acad. Sci. U. S. A. **109**, 15185 (2012)
8. A. Laurain, M. Scheller, P. Polynkin, Phys. Rev. Lett. **113**, 253901 (2014)
9. D. Kartashov et al., Phys. Rev. A **86**(3), 033831 (2012)
10. A. Dogariu, R.B. Miles, *In Frontiers in Optics 2013/Laser Science XXIX, Orlando, Florida, 2013* (Orlando, Laser Science, 2013)
11. V. Kocharovsky et al., Proc. Natl. Acad. Sci. U S A. **102**, 7806 (2005)
12. Q. Luo, W. Liu, S.L. Chin, Appl. Phys. B Lasers Opt. **76**, 337 (2003)
13. P.N. Malevich et al., Opt. Lett. **40**, 2469 (2015)
14. D. Kartashov et al., Phys. Rev. A **88**, 041805(R) (2013)
15. Y. Liu, Y. Brelet, G. Point, A. Houard, A. Mysyrowicz, Opt. Express **21**(19), 22791–22798 (2013)
16. P.J. Ding et al., Opt. Express **22**, 29964 (2014)
17. J. Yao et al., Opt. Express **22**, 19005 (2014)
18. S. Mitryukovskiy et al., Phys. Rev. Lett. **114**, 063003 (2015)
19. P.J. Ding, E. Oliva, A. Houard, A. Mysyrowicz, Y. Liu, Phys. Rev. A **94**, 043824 (2016)
20. T. Wang et al., Las. Phys. Lett. **10**, 125401 (2013)
21. S.L. Chin, H. Xu, Y. Cheng, Z. Xu, Chin. Opt. Lett. **10**, 013201 (2013)
22. D. Kartashov et al, Research in Optical Sciences, HTh4b. **5** (2016)
23. A. Azarm, P. Corkum, P. Polynkin, CLEO: Applications and Technology 2016, postdeadline paper JTh4B.9 (2016)
24. A. Couairon, A. Mysyrowicz, Phys. Rep. **441**, 47–198 (2007)
25. H.L. Xu, A. Azarm, J. Bernhardt, Y. Kamali, S.L. Chin, Chem. Phys. **360**, 171–175 (2009)
26. R. Arnold, S. Roberson, P.M. Pellegrino, Chem. Phys. **405**, 9 (2012)

27. A. Becker, A.D. Bandrauk, S.L. Chin, Chem. Phys. Let. **343**, 345 (2001)
28. P.H. Bucksbaum, M. Bashkansky, R.R. Freeman, T.J. McIlrath, L.F. DiMauro, Phys. Rev. Lett. **56**, 2590–2593 (1986)
29. P.B. Corkum, N.H. Burnett, F. Brunel, Phys. Rev. Lett. **62**, 1259–1262 (1989)
30. S. Mitryukovskiy, Y. Liu, A. Houard, A. Mysyrowicz, J. Phys. B Atomic Mol. Phys. **48**, 094003 (2015)
31. X.-L. Liu, W. Cheng, M. Petrarca, P. Polynkin, Opt. Lett. **41**, 4751 (2016)
32. R.S. Kunabenchi, M.R. Gorbal, M.I. Savadatti, Prog. Quant. Electron. **9**, 259 (1984)
33. H.M. von Bergmann, V. Hasson, J. Phys. D. Appl. Phys. **11**, 2341 (1978)
34. T. Zhao et al., J. Phys. D : Appl. Phys. **46**, 345201 (2013)
35. P. Sprangle, J. Peñano, B. Hafizi, D. Gordon, M. Scully, Appl. Phys. Lett. **98**, 211102 (2011)
36. J. Peñano et al., J. Appl. Phys. **111**, 033105 (2012)
37. E. Oliva et al., Phys. Rev. A **84**, 013811 (2011)
38. O. Larroche et al., Phys. Rev. A **62**, 043815 (2000)
39. T. Tabata, T. Shirai, M. Sataka, H. Kubo, At. Data Nucl. Data Tables **92**, 375–406 (2006)
40. R. Al'miev et al., Phys. Rev. Lett. **99**, 123902 (2007)
41. J. Yao et al., New J. Phys. **15**, 023046 (2013)
42. Y.D. Qin, H. Yang, C.J. Zhu, Q. Gong, Appl. Phys. B Lasers Opt. **71**, 581–584 (2000)
43. W. Chu et al., Las. Phys. Lett **11**, 015301 (2014)
44. J. Yao et al., Arxiv **1608**, 05183 (2016)
45. H. Zhang et al., Phys. Rev. **X**, 041009 (2013)
46. D. Kartashov et al, CLEO: science and innovations, QTh4E. **6** (2012)
47. G. Li et al., Phys. Rev. A **89**, 033833 (2014)
48. J.C. MacGillivray, M.S. Feld, Phys. Rev. A **14**, 1169 (1976)
49. H. Zhang et al., Phys. Rev. A **88**, 063417 (2013)
50. P.B. Corkum, Phys. Rev. Lett. **71**, 1994 (1993)
51. N.H. Burnett, C. Kan, P.B. Corkum, Phys. Rev. A **51**, R3418 (1995)
52. Y. Ivanov, T. Brabec, N. Burnett, Phys. Rev. A **54**, 742 (1996)
53. M. Kakehata, H. Takada, H. Yumoto, K. Miyazaki, Phys. Rev. A **55**, R861 (1997)

Chapter 4
Molecular Rotational Effects in Free-Space N_2^+ Lasers Induced by Strong-Field Ionization

Jinping Yao, Bin Zeng, Wei Chu, Haisu Zhang, Jielei Ni, Hongqiang Xie, Ziting Li, Chenrui Jing, Guihua Li, Huailaing Xu, and Ya Cheng

4.1 Introduction

Strong-field molecular physics is an important subject of contemporary physics that opens up fascinating opportunities for various applications ranging from molecular orbital imaging [1] and coherent X-ray sources [2] to attosecond chemistry [3], control of the long-range nonlinear propagation (laser filamentation) [4], and so forth. Interaction of molecules with intense laser fields has given rise to many intriguing physical effects, such as above-threshold ionization and dissociation [5], high-order harmonic generation [6], bond softening and hardening [7], molecular alignment [8], and interference of multiple orbitals [9]. Recently, it was discovered that the interaction of nitrogen molecules with ultrashort intense laser pulses can result in an instantaneous population inversion between the excited and ground electronic states of nitrogen molecular ions (N_2^+) [10–14]. This allows the generation of coherent narrow-bandwidth emissions in the forward direction in the presence of self-generated or externally injected seed pulses [10–35].

Surprisingly, the abovementioned externally seeded N_2^+ laser signals can be strongly affected by rotational wave packets of N_2^+ molecular ions in either the

J. Yao • B. Zeng • W. Chu • H. Zhang • J. Ni • H. Xie • Z. Li • C. Jing • G. Li
State Key Laboratory of High Field Laser Physics, Shanghai Institute of Optics and Fine Mechanics, Chinese Academy of Sciences, Shanghai 201800, China

H. Xu
State Key Laboratory on Integrated Optoelectronics, College of Electronic Science and Engineering, Jilin University, Changchun 130012, China

Y. Cheng (✉)
Shanghai Institute of Optics and Fine Mechanics, East China Normal University, Shanghai, China
e-mail: ya.cheng@siom.ac.cn

P. Polynkin, Y. Cheng (eds.), *Air Lasing*, Springer Series in Optical Sciences 208, https://doi.org/10.1007/978-3-319-65220-7_4

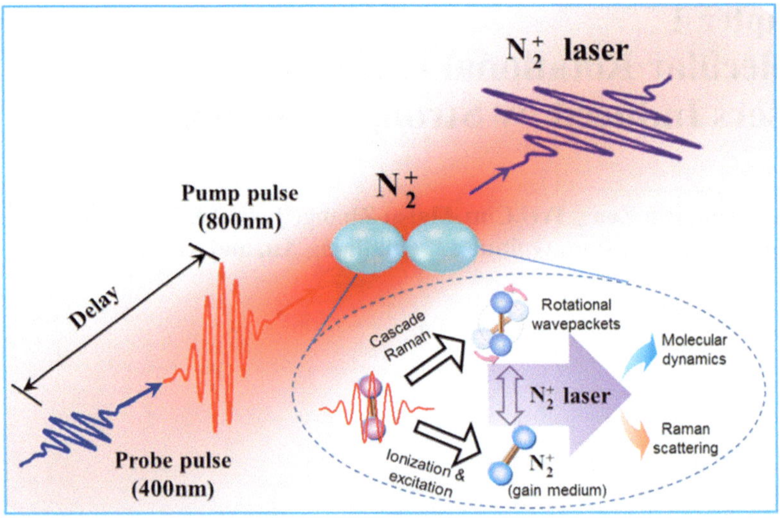

Fig. 4.1 Schematic diagram of the pump-probe experimental setup for the investigation of the rotational effects in the free-space N_2^+ laser. In the pump-probe scheme, the 800 nm pulse is used as a pump to generate the N_2^+ ions in the population-inverted state; meanwhile, the coherent rotational wave packets of both neutral N_2 molecules and N_2^+ ions are also created through the non-resonant cascade Raman process by the pump pulse. After a time delay, the second harmonic of the pump laser is injected into the population-inverted N_2^+ ions, which serves as a probe of the stimulated amplification at either 391 or 428 nm wavelength. The rotational coherence created by the pump pulse can faithfully encode its characteristics into the amplified probe signal.

ground $X^2\Sigma_g^+(v=0)$ state or the excited $B^2\Sigma_u^+(v'=0)$ state [11, 26, 28]. Investigations on the rotational effects with a pump-probe scheme, as schematically illustrated in Fig. 4.1, not only provide crucial clues for understanding the mechanism behind the population inversion in the N_2^+ ions but also open new possibilities for the control of the rotational wave packets by tailoring the driver field. Experimentally, by measuring the evolution of the laser signal with the increasing time delay between the pump and probe pulses, one can probe the ultrafast dynamics of the coherent rotational wave packets in the N_2^+ ions [11, 26, 28]. These measurements can be performed with high temporal resolution, determined by the temporal durations of the pump and probe pulses, as well as with high spectral resolution, determined by the spectral width of the generated N_2^+ lasers.

In this chapter, we will review the experimental observations made over the past few years that highlight the molecular rotational effects in the strong-field-ionization-induced N_2^+ lasers. The chapter is organized as follows: First, we will describe the real-time observation of the ultrafast dynamics of the coherent rotatio nal wave packets using the N_2^+ laser as a probe and show how the N_2^+ lasers can be influenced by coherent coupling of the rotational quantum states in the strong laser field. Then, we will show that the polarization state of the N_2^+ laser can be manipulated through the rotational effects. We will also demonstrate the impulsive Raman scattering in neutral nitrogen molecules using the N_2^+ laser as a probe source. We will conclude with a summary of the results discussed in the chapter and a perspective on future applications.

4.2 Real-Time Observation of the Ultrafast Dynamics of the Coherent Rotational Wave Packets

A typical N_2^+ laser spectrum generated in a pure nitrogen gas at 20 mbar pressure is shown in Fig. 4.2a [11]. It can be clearly seen that a strong, narrow-linewidth emission appears at 391.4 nm, which corresponds to the P branch of the $B^2\Sigma_u^+(v'=0) \rightarrow X^2\Sigma_g^+(v=0)$ transition in molecular nitrogen ions. It can also be noted that a series of separated peaks appear on the blue side of the P-branch laser line, which are the transitions within the R-branch band of the same electronic state. Each peak corresponds to the stimulated emission from a specific rotational energy level J of the $B^2\Sigma_u^+(v'=0)$ state to the rotational energy level (J-1) of the $X^2\Sigma_g^+(v=0)$ state, as illustrated in Fig. 4.2b.

As mentioned in Sect. 1, the free-space N_2^+ laser allows us to reveal the ultrafast dynamics of coherent rotational wave packets in a strong laser field through pump-probe measurements. Generally speaking, alignment of N_2^+ ions can be achieved either by an impulsive Raman excitation [8] or by the preferential ionization of the N_2 molecules with the molecular axes parallel to the laser polarization [36] or by both of them. With the alignment, the rotational coherence of the wave packets would survive after the pump pulse is turned off. The free evolution of the coherent rotational wave packets causes the rotational eigenstates to dephase and rephase periodically. To verify the rotational coherence, Zhang et al. measured the signal intensity of the P-branch bandhead at ~391 nm as a function of the time delay between the pump and probe pulses for the polarization directions of the two pulses either parallel or perpendicular to each other [11]. As shown in Fig. 4.3a, both curves show a rapid increase followed by a slow exponential decay with the decay time of 3–4 ps. The measurement reveals the gain dynamics of N_2^+ lasers induced by strong-field ionization, i.e., a fast growth in the beginning followed by a slow decay. Moreover, periodic modulations appear on the measured gain curves. The periodic modulations of the signal at approximately 2.0 ps, 4.0 ps, 6.0 ps, 8.0 ps, and 10.0 ps can be reasonably assigned to $T_{rot}/4$, $T_{rot}/2$, $3T_{rot}/4$, T_{rot}, and $5T_{rot}/4$ revivals of the rotational wave packet of the $B^2\Sigma_u^+(v'=0)$ state of N_2^+, which has a rotational period T_{rot} = 8.0 ps. A similar result can be obtained if we plot the overall signal in the R-branch band as a function of the time delay, as shown in Fig. 4.3b.

To better understand the correlation between the measured results in the time and frequency domains, we perform a Fourier transform for the two curves in Fig. 4.3a after removing the exponentially decaying baselines. The corresponding Fourier spectra for the cases of parallel and perpendicular polarization are presented in

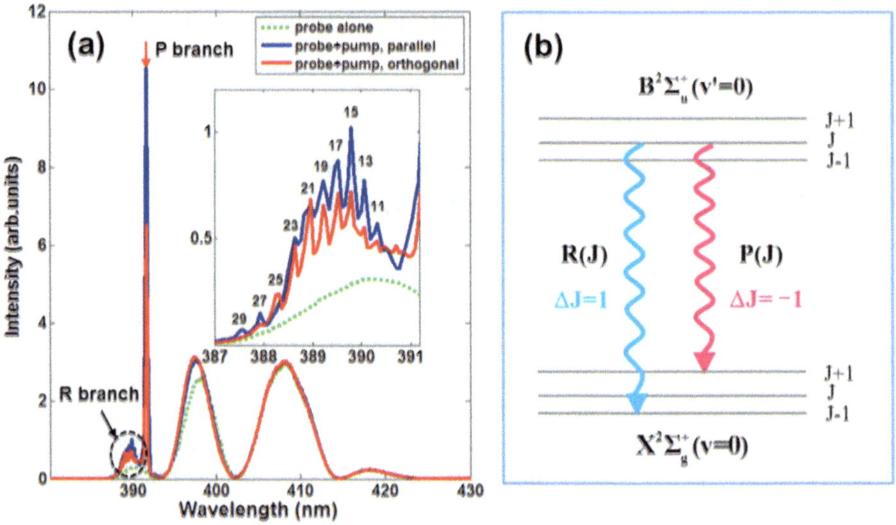

Fig. 4.2 (**a**) A typical spectrum of a forward-propagating amplified seed at low pressure. The inset shows the spectrum in the range 387–391 nm, with numbered rotational levels of the upper emission state of N_2^+. For comparison, the original seed spectrum is shown with the *green dotted line*. (**b**) Schematic diagram of the rotational transitions between $B^2\Sigma_u^+(v'=0)$ and $X^2\Sigma_g^+(v=0)$ states [11]

Fig. 4.3c, d, respectively. Clearly, the two spectra show similar frequency distributions with several peaks corresponding to the beat frequencies between the odd J and $J+2$ states. The frequency peak with the largest amplitude belongs to the beat frequency between the $J = 13$ and $J = 15$ rotational states. The Fourier spectrum is in good agreement with the measured rotational-state distribution in the R-branch spectrum shown in Fig. 4.2a. It is noteworthy that weak even-order J peaks can only be observed in Figs. 4.3c, d, whereas they could not be observed in Fig. 4.2a, owing to the fact that the spectral resolution in the time-domain measurement is higher than the spectral resolution in the frequency-domain measurement.

The experimental results shown in Fig. 4.3a can be theoretically reproduced as follows: The rotational wave packet of the N_2^+ ions left behind the pump laser pulses can be written as $\psi_0 = \sum a_J \mid J, M\rangle$. Since the transition from the $B^2\Sigma_u^+$ state to the $X^2\Sigma_g^+$ state is a parallel transition, the strength of the stimulated emission from the N_2^+ ions aligned parallel to the polarization direction of the probe pulse is stronger than that from those aligned perpendicularly to it. Consequently, the probe pulses injected into the plasma spark with different delay times see different alignment angles of N_2^+ ions and thus experience different amplification efficiencies. Thus, for the parallel polarization of the pump and probe pulses, the time-dependent

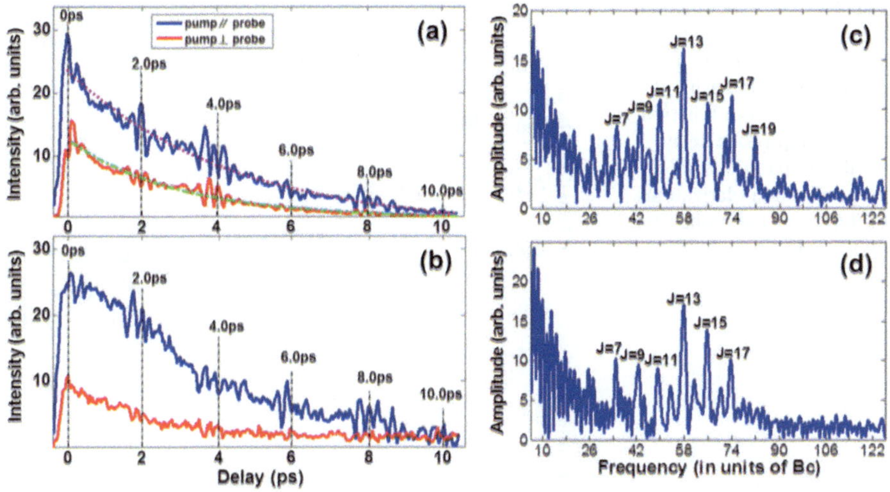

Fig. 4.3 Signal intensities of (**a**) P-branch bandhead at around 391 nm and (**b**) R-branch band recorded over the wavelength range 387–390.7 nm as functions of the time delay between the pump and probe pulses. The dotted lines are exponential fits to the experimental curves. The vertical lines indicate the rotational revival times of the $B^2\Sigma_u^+$ state. The Fourier transforms of the oscillation curves shown in the panel (**a**) for the cases of (**c**) parallel and (**d**) perpendicular polarization directions of the pump and probe pulses. The numbers indicate the rotational quantum numbers J in the upper energy state [11]

probability of stimulated emission in the ensemble of aligned N_2^+ ions can be written as

$$\langle\cos^2\theta\rangle(t) = \sum_J |a_J|^2 C_{J,J,M=0} + |a_J||a_{J+2}|\cos\left(\Delta\omega_{J,J+2}t + \phi_{J,J+2}\right)C_{J,J+2,M=0} \tag{4.1}$$

where θ is the angle between the molecular axis and the polarization of the probe pulse and $|a_J|$ and $|a_{J+2}|$ the probability amplitudes of $|J, M\rangle$ and $|J+2, M\rangle$ states, respectively; $\Delta\omega_{J,J+2}$ the beat frequency between $|J, M\rangle$ and $|J+2, M\rangle$ states, $\phi_{J,J+2}$ the relative phase of the two states at the beginning of free evolution, and $C_{J,J,M=0} = \langle J, M|\cos^2\theta| J, M\rangle$ and $C_{J,J+2,M=0} = \langle J, M|\cos^2\theta| J+2, M\rangle$ are constants. Here, $|a_J|$ was approximately evaluated by the signal intensities of the R-branch band of the laser spectrum in Fig. 4.2a, and the evolution of the rotational wave packet was calculated from the moment when the initial highest alignment degree was achieved after the pump pulse. At the moment, each rotational eigenstate had approximately the same initial phase, namely, $\phi_{J,J+2} = 0$. Therefore, the variation of the signal intensities at different time delay, τ, can be expressed by the function of $\exp(-\tau/\Gamma) \times \langle\cos^2\theta\rangle(\tau)$. The exponential term describes the decay of population inversion between $B^2\Sigma_u^+$ and

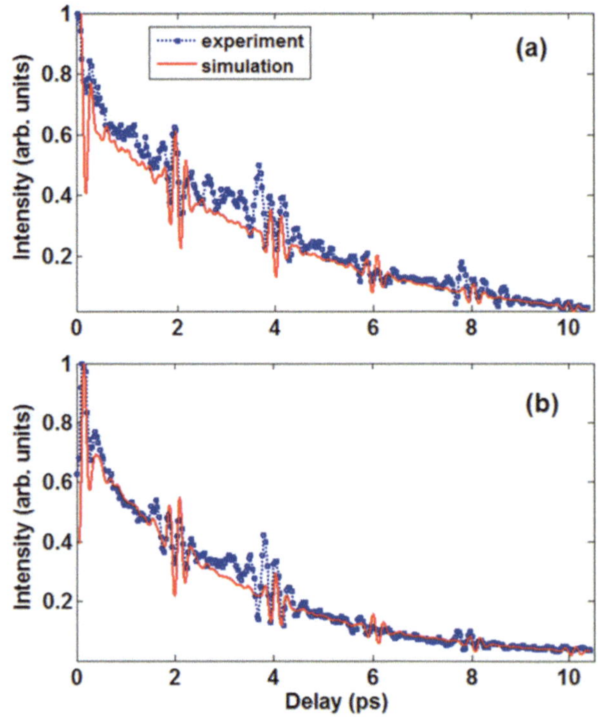

Fig. 4.4 The experimental (*blue dotted lines*) and simulated (*red solid lines*) curves for the intensity of the N_2^+ laser as a function of the pump-probe delay for the (**a**) parallel and (**b**) perpendicular polarizations of the pump and probe pulses [11]

$X^2\Sigma_g^+$ states. The simulated results for the pump and probe pulses in the parallel and orthogonal polarization cases are plotted in Figs. 4.4a, b, respectively. It can be seen that the major features of the experimental curves (blue dotted curves) are qualitatively reproduced by the simulation (red solid curves), which confirms the buildup of the coherent rotational wave packet of the N_2^+ ions in the plasma spark produced by the intense femtosecond pump pulse.

4.3 Coherent Coupling of the Rotational Quantum States in the Strong Laser Field

Time-resolved spectrosopic measurements with high resolution allow us to investigate coherent coupling of rotational quantum states in a strong laser field [26, 28]. As an example, Fig. 4.5a shows the intensities of several R-branch lasing lines from the individual rotational states with different quantum numbers *J* as a function of the time delay between the pump and probe pulses [28]. In this experiment, the lasing signal is generated by collinearly focusing a 2 mJ, 800 nm pump pulse and a time-delayed 400 nm probe pulse into the pure nitrogen gas at 3 mbar pressure. Unlike the P-branch transition shown in Fig. 4.3a, the temporal evolution of the R-branch transition from a specific rotational state shows fast oscillations with a

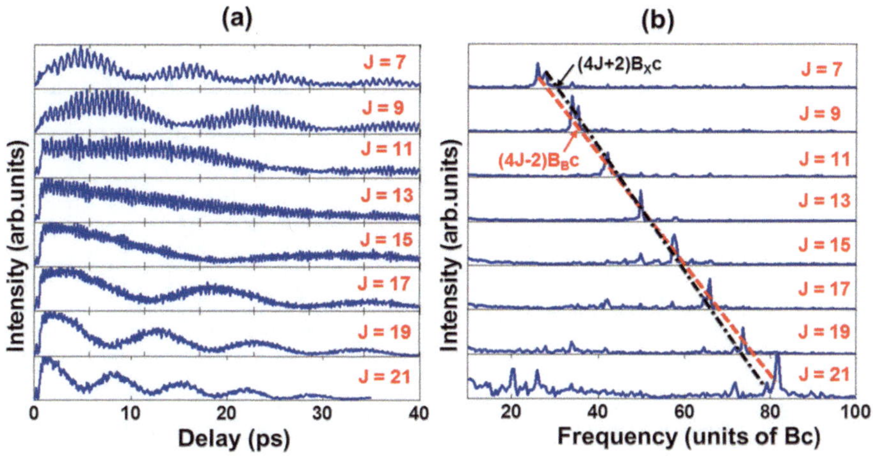

Fig. 4.5 (**a**) Intensities of the R-branch lasing lines corresponding to the transitions $B^2\Sigma_u^+(v'=0,J)\rightarrow X^2\Sigma_g^+(v\ =0,J-1)$ of the N_2^+ ions, as functions of the time delay between the pump and probe pulses at the pump energy of 2 mJ. (**b**) The Fourier transforms of the corresponding curves in (**a**) [28]

sub-picosecond period, together with a slow oscillation with multi-picosecond period. For the fast oscillation, its period becomes shorter with the increase of the rotational quantum number *J*. For the slow oscillation, its period first increases with the rotational quantum number *J* until the quantum number reaches *J* = 13; then it decreases as the rotational quantum number further increases. Interestingly, for the rotational number *J* = 13, the slow oscillation disappears.

To understand the origin of intensity oscillations of different emission lines, Fourier transforms were applied to these time-dependent curves. As shown in Fig. 4.5b, the Fourier transformed spectrum of the fast oscillations contains one strong frequency component along with one weak frequency component. As indicated by the red dashed line in Fig. 4.5b, the strong frequency component can be fitted with the beat frequency between the $|J\rangle$ and $|J-2\rangle$ states of $N_2^+(B^2\Sigma_u^+)$, which is written as $\Delta\omega_{J,J-2}^{B}=(4J-2)B_Bc$. Here, the value of the rotational constant is B_B = 2.073cm^{-1}, and *c* is the speed of light. The weak frequency component can be well fitted by the beat frequency of the $|J-1\rangle$ and $|J+1\rangle$ states of $N_2^+(X^2\Sigma_g^+)$, which is written as $\Delta\omega_{J-1,J+1}^{X}=(4J+2)B_Xc$, as indicated by the black dash-dotted line. Here, the rotational constant is B_X = 1.92cm^{-1}.

Furthermore, when pump energy is increased from 2 to 2.4 mJ, the R-branch components of the N_2^+ laser show a broad distribution of rotational states with the highest rotational quantum number *J* = 29, as shown in Fig. 4.6a. As revealed by the Fourier transform spectra shown in Fig. 4.6b, in addition to the previously observed frequency component at $(4J-2)B_Bc$, more beat frequencies appear in the case of high-intensity pump laser, such as frequencies $(4J-10)B_Bc$ and $(4J+6)B_Bc$. Surprisingly, as indicated by the black dash-dotted line, an unexpected frequency component appears in the Fourier transform spectra shown in Fig. 4.6b, of which the Fourier frequency decreases linearly with the increasing value of *J*. The red

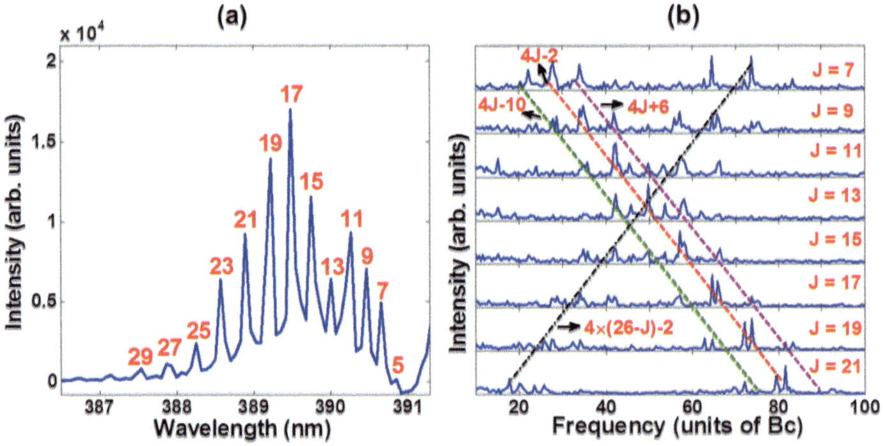

Fig. 4.6 (**a**) Spectrum of the forward-propagating lasing on the R-branch transitions at the pump energy of 2.4 mJ. (**b**) The Fourier transform spectra of the time-dependent lasing intensities of the R-branch transitions [28]

dashed and black dash-dotted lines form an "X" structure in the Fourier transform spectra. The physical origin of all of these frequency components in the Fourier spectra of the lasing lines can be attributed to the coupling between different rotational energy levels of the $N_2^+(B^2\Sigma_u^+)$ and $N_2^+(X^2\Sigma_g^+)$ states mediated by the exchange of the R-branch and P-branch photons in the laser spectrum, as revealed by the theoretical analyses of Ref. [28]. Thus, the dependence of the frequency components on the rotational quantum number J reflects the redistribution of populations in different rotational energy levels due to the interaction of multiple rotational wave packets with the ultrafast laser pulse.

4.4 Modulation of Polarization of the N_2^+ Laser Due to Molecular Rotations

Strong birefringence was observed in the population-inverted N_2^+ medium, which enabled the manipulation of polarization of stimulated emission from the coherent rotational wave packets [34]. When the angle θ between the polarization directions of the pump and seed pulses varies in the range of 0–90°, the N_2^+ laser becomes elliptically polarized with variable ellipticity. Moreover, the polarizations of the P-branch and R-branch lines in the N_2^+ laser emission show significantly different behaviors.

Figure 4.7 shows intensities of the P-branch (red diamonds) and R-branch (blue circles) components of the N_2^+ lasing, as well as the intensity of the seed (black stars), as functions of the angle of the polarizer placed before the spectrometer. Here, the angle θ indicated in each panel is changed by varying the polarization direction

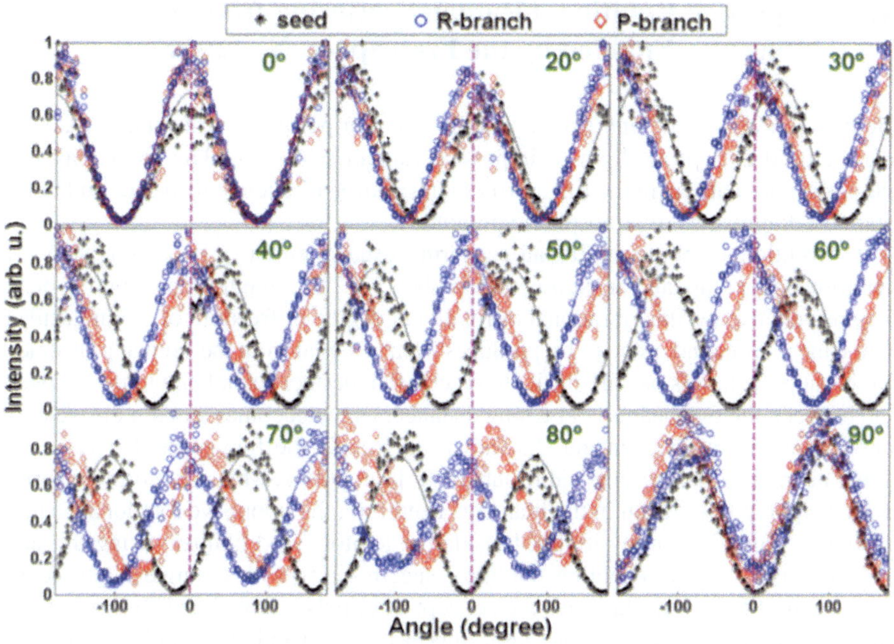

Fig. 4.7 Measured intensities of the P-branch and R-branch lines of the N_2^+ laser, as well as that of the seed pulse, as a function of the angle of the polarizer placed before the spectrometer. The angle θ between the polarization directions of the pump and seed pulses is indicated in each panel [34]

of the seed pulse, while the polarization direction of the pump is fixed. Pressure of the nitrogen gas is 13.5 mbar, and the time delay between the pump and seed pulses is set to 1.5 ps, which is significantly different from the time of molecular alignment revival for nitrogen. It is observed that when the angle θ is less than 70°, the N_2^+ laser is nearly linearly polarized, and its polarization direction is almost parallel to that of the pump pulse but not to that of the seed pulse. As the angle θ increased to 80°, the N_2^+ laser emission becomes significantly elliptically polarized, with the ellipticity of 0.28. The P-branch and R-branch emissions have different polarization directions in spite of the fact that their wavelengths are very close to each other. When the angle θ is tuned to 90°, the N_2^+ laser becomes linearly polarized, with its polarization direction parallel to that of the seed pulse, which is consistent with the previous observations. This variation of the polarization state of N_2^+ lasers could be qualitatively explained by the strong birefringence of the gain medium near the wavelengths of the N_2^+ laser. The similar results were obtained at different gas pressures, pump energies, and pump-probe delays, indicating that the observed phenomenon is universal.

4.5 Impulsive Rotational Raman Scattering from N_2 Molecules Enabled by the Free-Space N_2^+ Laser

Being a source of coherent, narrow-bandwidth laser pulses, the free-space N_2^+ laser can be used as a probe light of impulsive Raman scattering. The latter may find potential applications in the fingerprint identification of chemical species and in the standoff diagnostics of atmospheric pollutants.

Conventional impulsive Raman scattering usually employs two laser beams, a femtosecond pump and a picosecond probe [37]. The femtosecond pump pulse impulsively excites the vibrational or rotational degree of freedom of the molecules, while the picosecond probe pulse produces Raman signal from the excited molecules. In this approach, it is usually difficult to achieve a perfect spatial and temporal overlap between the two beams over a long distance, which becomes a major obstacle for applications in remote sensing. As shown in Sect. 2, the free-space N_2^+ laser can generate intense, narrow-linewidth, sub-10-ps laser pulses, which are co-propagating with the pump laser beam in the filament. Therefore, the impulsive rotational Raman scattering can be remotely initiated inside the femtosecond filament using the self-induced lasing signal, enabling nonlinear spectroscopy with a single femtosecond laser beam. Figure 4.8 illustrates the concept of impulsive rotational Raman scattering using an intense femtosecond laser pulse as a pump and the stimulated emission from the N_2^+ ions, excited by the pump laser, as a probe. Raman spectroscopy based on the free-space N_2^+ lasing has two unique advantages. First, the pump and the probe pulses naturally overlap in both time and space, since the probe pulse is generated inside the filament induced by the pump laser. Second, femtosecond laser filamentation in air has a potential to realize Raman fingerprinting at a distance.

Ni et al. had first observed impulsive Raman scattering of N_2 molecules with the free-space N_2^+ laser driven by an 800 nm, 40 fs, 15 mJ pump laser [25]. The experimental setup is schematically shown in Fig. 4.9a. The high-energy laser beam is first focused into the nitrogen gas at 1 atmosphere pressure by the lens with a focal length of 40 cm, to form a filament. The forward-propagating signal is collimated by the lens with a focal length of 40 cm, after reflection by an uncoated glass plate. The lasing emission around 428 nm is isolated from the supercontinuum by an interference filter with the central wavelength of 428 nm and the bandwidth of 10 nm. Finally, the signal is focused onto the entrance slit of an imaging spectrometer by a lens. The inset in Fig. 4.9a shows a strong, narrow-linewidth emission centered at 427.8 nm, which is attributed to the self-seeded lasing on the transition between the $B^2\Sigma_u^+(v'=0)$ and $X^2\Sigma_g^+(v=0)$ states of the N_2^+ ions. The seed is provided by the supercontinuum white light generated in the femtosecond filament. Besides the strong laser line at 427.8 nm, several closely spaced peaks can also be observed on both sides of the line. Calculation shows that the peaks on the blue side of the 427.8 nm lasing line are the mixture of the R-branch rotational transitions in N_2^+ and the O-branch Raman scattering from neutral nitrogen molecules. The peaks on the red side correspond to the S-branch Raman scattering from neutral nitrogen molecules.

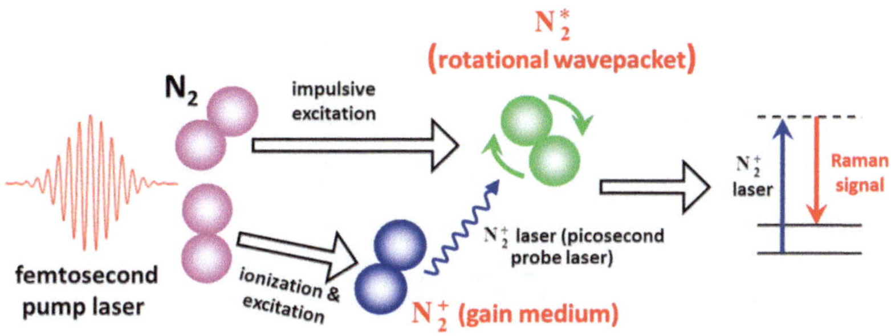

Fig. 4.8 Schematic diagram of impulsive Raman scattering induced by the free-space N_2^+ laser.

Next, we focus on the S-branch Raman scattering induced by the free-space N_2^+ laser. For clarity, the spectrum shown in the inset in Fig. 4.9a is replotted in wavenumber units using the unit conversion $10^7/\lambda - 10^7/427.8$, where λ is the wavelength of the lasing signal. As shown in Fig. 4.9b, the peaks on the red side of the lasing spectrum are well separated, and the frequency difference between the two successive peaks is nearly constant, with the value of ~16 cm^{-1}. For the S-branch rotational Raman scattering shown in Fig. 4.9c, the spectral shift is $\Delta\nu = B(4J + 6)$, where $B = 1.998\text{cm}^{-1}$ is the rotational constant of the $N_2\left(X^1\Sigma_g^+\right)$ state and J is the rotational quantum number. The calculated wavenumber shifts of the rotational Raman lines in the S-branch ($J \rightarrow J + 2$), S(6), S(8), S(10), S(12), and S(14), are 59.94 cm^{-1}, 75.92 cm^{-1}, 91.91 cm^{-1}, 107.89 cm^{-1}, and 123.88 cm^{-1}, respectively, as shown by the red circles. These values are in good agreement with the measured spectral shifts of 59.8 cm^{-1}, 76.6 cm^{-1}, 91.0 cm^{-1}, 107.5 cm^{-1}, and 124.4 cm^{-1}. These calculations unambiguously assign the measured spectral peaks to the S-branch transitions of the impulsive rotational Raman scattering from neutral nitrogen molecules. For the strongest line S(6), the conversion efficiency of Raman scattering is estimated to be about 0.8%, by dividing the intensity of the Raman signal by the intensity of the lasing line at 427.8 nm. Thus, impulsive Raman scattering enabled by air lasing provides a promising way for remote identification of chemical species in the atmosphere.

4.6 Summary and Future Perspective

In summary, we have reviewed the experimental observations of rotational effects in the strong-field-induced N_2^+ lasing. By pump-probe measurements, the rotational spectrum of the N_2^+ laser can be examined, with high spectral and temporal resolutions. In the frequency domain, the linewidth of the strong-field-ionization-induced N_2^+ laser is typically on the order of 0.1 nm, which allows for the retrieval of the rotational state distribution of the N_2^+ ions from the R-branch spectra. Pulse durations

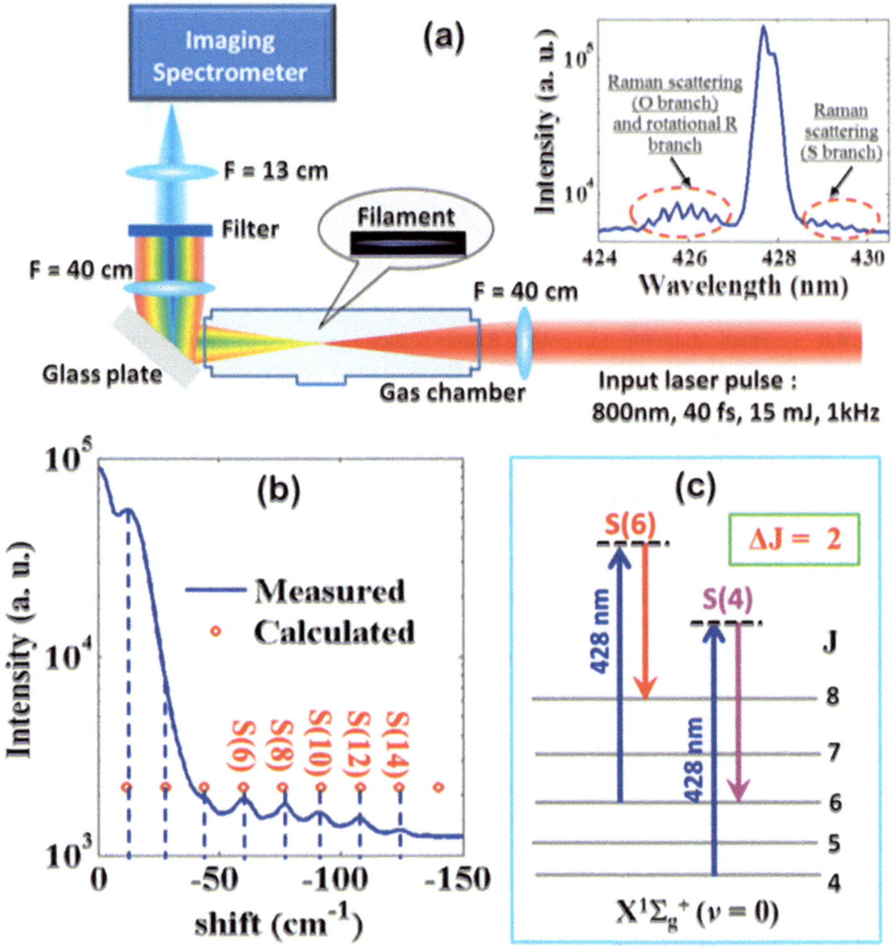

Fig. 4.9 (**a**) Schematic diagram of the experimental setup for impulsive Raman scattering based on the free-space N_2^+ laser. *Inset*: measured lasing spectrum in the forward direction. (**b**) Sbranch of the measured Raman spectrum (*blue solid line*) and the calculated Raman lines in nitrogen (*red circles*). (**c**) Schematic diagram of the S-branch transition of Raman scattering from nitrogen molecules [25]

of the pump and probe lasers are both on the order of 100 fs or less, enabling high-resolution measurements of the ultrafast dynamics of coherent rotational wave packets in time domain. When operating in the mode of self-generated seed amplification, the high-intensity, narrow-linewidth N_2^+ laser can be an ideal source for generating impulsive Raman scattering, which may become a useful tool for the standoff detection of atmospheric pollutants.

Current investigations mainly focus on the rotational effects. In principle, the same pump-probe approach can be extended to the investigations on vibrational dynamics, by using probe pulses of durations approaching the vibrational periods of the molecular ions. This technique holds promise for both ultrafast spectroscopy and remote sensing, thanks to the unique characteristics of air lasers such as the high degree of coherence, high brightness, directionality, narrow linewidth, and switchable wavelength.

References

1. J. Itatani et al., Tomographic imaging of molecular orbitals. Nature **432**, 867 (2004)
2. C. Vozzi et al., Controlling two-center interference in molecular high harmonic generation. Phys. Rev. Lett. **95**, 153902 (2005)
3. H.J. Worner et al., Conical intersection dynamics in NO_2 probed by homodyne high-harmonic spectroscopy. Science **33**, 208 (2011)
4. S. Varma, Y.H. Chen, H.M. Milchberg, Trapping and destruction of long-range high-intensity optical filaments by molecular quantum wakes in air. Phys. Rev. Lett. **101**, 205001 (2008)
5. A. Giusti-Suzor et al., Above-threshold dissociation of H_2^+ in intense laser fields. Phys. Rev. Lett. **64**, 515 (1990)
6. R. Vellota et al., High-order harmonic generation in aligned molecules. Phys. Rev. Lett. **87**, 183901 (2001)
7. P.H. Bucksbaum et al., Softening of the H_2^+ molecular bond in intense laser fields. Phys. Rev. Lett. **64**, 1883 (1990)
8. T. Seideman, Rotational excitation and molecular alignment in intense laser fields. J. Chem. Phys. **103**, 7887 (1995)
9. O. Smirnova et al., High harmonic interferometry of multi-electron dynamics in molecules. Nature **460**, 972 (2009)
10. J.P. Yao et al., Remote creation of coherent emissions in air with two-color ultrafast laser pulses. New J. Phys. **15**, 023046 (2013)
11. H.S. Zhang et al., Rotational coherence encoded in an "air-laser" spectrum of nitrogen molecular ions in an intense laser field. Phys. Rev. X **3**, 041009 (2013)
12. J.L. Ni et al., Identification of the physical mechanism of generation of coherent N_2^+ emissions in air by femtosecond laser excitation. Opt. Express **21**, 8746 (2013)
13. H.L. Xu et al., Sub-10-fs population inversion in N_2^+ in air lasing through multiple state coupling. Nat. Commun. **6**, 8347 (2015)
14. J.P. Yao et al., Population redistribution among multiple electronic states of molecular nitrogen ions in strong laser fields. Phys. Rev. Lett. **116**, 143007 (2016)
15. J.P. Yao et al., High-brightness switchable multiwavelength remote laser in air. Phys. Rev. A **84**, 051802(R) (2011)
16. D. Karatashov et al., Stimulated Amplification of UV Emission in a Femtosecond Filament Using Adaptive Control, in *Conference on Lasers and Electro-Optics*, OSA Technical Digest (Optical Society of America, 2012), paper QTh4E.6 (2012)
17. Y. Liu et al., Self-seeded lasing in ionized air pumped by 800 nm femtosecond laser pulses. Opt. Express **21**, 22791 (2013)
18. T.J. Wang et al., Self-seeded forward lasing action from a femtosecond Ti:Sapphire laser filament in air. Laser Phys. Lett. **10**, 125401 (2013)
19. W. Chu et al., A self-induced white light seeding laser in a femtosecond laser filament. Laser Phys. Lett. **11**, 015301 (2013)
20. S.L. Chin et al., Natural population inversion in a gaseous molecular filament. Chin. Opt. Lett. **11**, 013201 (2013)
21. G. Andriukaitis et al., Intense, directional UV emission from molecular nitrogen ions in an adaptively controlled femtosecond filament. EPJ Web of Conferences **41**, 10004 (2013)

22. H.S. Zhang et al., Abnormal dependence of strong-field-ionization-induced nitrogen lasing on polarization ellipticity of the driving field. Phys. Rev. A **88**, 063417 (2013)
23. T.J. Wang et al., Forward lasing action at multiple wavelengths seeded by white light from a femtosecond laser filament in air. Phys. Rev. A **88**, 053429 (2013)
24. A. Baltuška, D. Kartashov, Transient Inversion in Rotationally Aligned Nitrogen Ions in a Femtosecond Filament, in *Research in Optical Sciences*, OSA Technical Digest (online) (Optical Society of America, 2014), paper HTh4B.5 (2014)
25. J.L. Ni et al., Impulsive rotational Raman scattering of N_2 by a remote "air laser" in femtosecond laser filament. Opt. Lett. **39**, 2250 (2014)
26. B. Zeng et al., Real-time observation of dynamics in rotational molecular wave packets by use of air-laser spectroscopy. Phys. Rev. A **89**, 042508 (2014)
27. G. Point et al., Lasing of ambient air with microjoule pulse energy pumped by a multi-terawatt infrared femtosecond laser. Opt. Lett. **39**, 1725 (2014)
28. H.Q. Xie et al., Coupling of N_2^+ rotational states in an air laser from tunnel-ionized nitrogen molecules. Phys. Rev. A **90**, 042504 (2014)
29. G.H. Li et al., Signature of superradiance from a nitrogen-gas plasma channel produced by strong-field ionization. Phys. Rev. A **89**, 033833 (2014)
30. C.R. Jing et al., Generation of an air laser at extended distances by femtosecond laser filamentation with telescope optics. Opt. Express **22**, 3151 (2014)
31. Y. Liu et al., Recollision-induced superradiance of ionized nitrogen molecules. Phys. Rev. Lett. **115**, 133203 (2015)
32. C.R. Jing et al., Dynamic wavelength switching of a remote nitrogen or air laser with chirped femtosecond laser pulses. Laser Phys. Lett. **12**, 015301 (2015)
33. P. Wang, C.Y. Wu, M.W. Lei, Population dynamics of molecular nitrogen initiated by intense femtosecond laser pulses. Phys. Rev. A **92**, 063412 (2015)
34. Z.T. Li et al., Generation of elliptically polarized nitrogen ion laser fields using two color femtosecond laser pulses. Sci. Rep. **6**, 21504 (2016)
35. A. Azarm, P. B. Corkum, P. G. Polynkin, Rotational Mechanism of Lasing in Singly Ionized Nitrogen Molecules under Femtosecond mid-IR Pumping, in *Conference on Lasers and Electro-Optics*, OSA Technical Digest (online) (Optical Society of America, 2016), paper JTh4B.9 (2016)
36. M. Spanner et al., Mechanisms of two-color laser-induced field-free molecular orientation. Phys. Rev. Lett. **109**, 113001 (2012)
37. A. Nazarkin et al., Generation of multiple phase-locked Stokes and anti-Stokes components in an impulsively excited Raman medium. Phys. Rev. Lett. **83**, 2560 (1999)

Chapter 5
Filament-Initiated Lasing in Neutral Molecular Nitrogen

Daniil Kartashov, Mikhail N. Shneider, and Andrius Baltuska

5.1 Introduction

Modern optical spectroscopy of the atmosphere at high altitudes relies primarily on the powerful and well-established LIDAR technique. A ground-based (or air–/space-borne) laser source is used for standoff linear measurements of the scattering characteristics of molecular, atomic, or particle species in the air in different scattering regimes (Rayleigh, Mie, spontaneous Raman). The incoherent nature of the scattered field measured in the backward direction sets practical limitation on the tracing distance, spatial resolution, and the species-dependent sensitivity threshold. The possibility of standoff initiation of a coherent source of backward-directed radiation in the sky would enable application of different methods of nonlinear optics, such as stimulated Raman scattering, CARS, etc., for the highly sensitive and highly selective spectroscopy of the atmosphere at high altitudes [1, 2] with (potentially) significantly larger detection range and finer spatial resolution. One of the possibilities to initiate such standoff coherent source of radiation in the atmosphere can be provided by achieving population inversion and laser generation in one of the two main atmospheric constituents, nitrogen ($\approx$78% concentration) and oxygen ($\approx$21% concentration).

The phenomenon of femtosecond laser filamentation in transparent media and, particularly, in air [3] opens very attractive perspectives for remote plasma generation and standoff lasing in the atmosphere. Femtosecond filamentation intrinsically

D. Kartashov (✉)
Friedrich-Schiller University Jena, Max-Wien-Platz 1, 07743 Jena, Germany
e-mail: daniil.kartashov@uni-jena.de

M.N. Shneider
Department of Mechanical and Aerospace Engineering, Princeton University, Princeton, NJ 08544-5263, USA

A. Baltuska
Photonics Institute, Vienna University of Technology, Gusshausstrasse 27-387, A-1040 Vienna, Austria

P. Polynkin, Y. Cheng (eds.), *Air Lasing*, Springer Series in Optical Sciences 208, https://doi.org/10.1007/978-3-319-65220-7_5

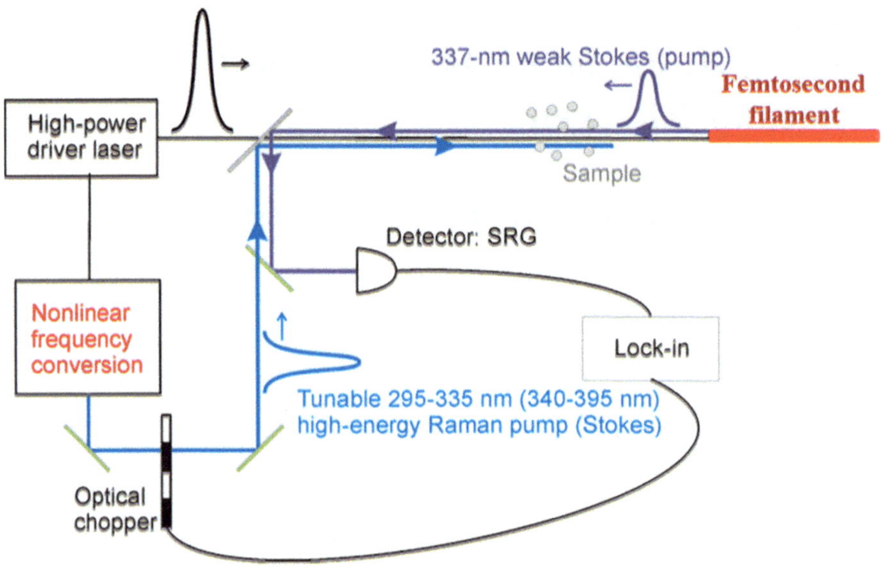

Fig. 5.1 A sketch of a setup for nonlinear spectroscopy of the atmosphere based on stimulated Raman gain/loss measurements

involves ionization as a stopping mechanism for the beam collapse due to self-focusing [4]. As a result, it enables standoff generation of plasma channels in air along the propagation path at the distances of up to 1 km [5] with plasma densities varying in a broad range 10^{14}–10^{18} cm^{-3} [6]. Such plasma channels are prerequisites for the realization of a cavity-free, single pass molecular nitrogen laser in the sky. A sketch of the possible realization of the nonlinear optical spectroscopy of the atmosphere using filament-initiated lasing is shown in Fig. 5.1. A high-power femtosecond laser pulse generates a filament or a multifilament bundle at a controllable distance. The plasma channel formed by the filament (or the filament bundle) enables a backward-propagating pulse of coherent UV radiation generated due to the single-pass lasing in molecular nitrogen. A portion of the high-power laser radiation can be used for a parametric frequency conversion, providing a synchronized tunable narrowband pulse that meets the UV pulse generated by the "sky" laser at a certain point, enabling stimulated Raman scattering. Depending on the difference between the frequency of the tunable source and the fixed frequency of the "sky" laser, nonlinear spectroscopy of air at the selected point in the atmosphere can be realized through stimulated Raman gain or stimulated Raman loss.

Laser generation in molecular nitrogen was demonstrated at the very dawn of the laser era [7]. Lasing occurs between the excited $C^3\Pi_u$ (C state) and $B^3\Pi_g$ (B state) electronic states and their vibrational manifolds (second positive system), resulting in the UV emission in the 300–400 nm spectral range, or between the $B^3\Pi_g$ and $A^3\Sigma_u$ (A state) states (first positive system), resulting in the near-IR emission band 0.75–1.23 μm (Fig. 5.2). It is worth mentioning that the fluorescence from the latter emission band strongly contributes to the infrared background radiation of the night sky [8].

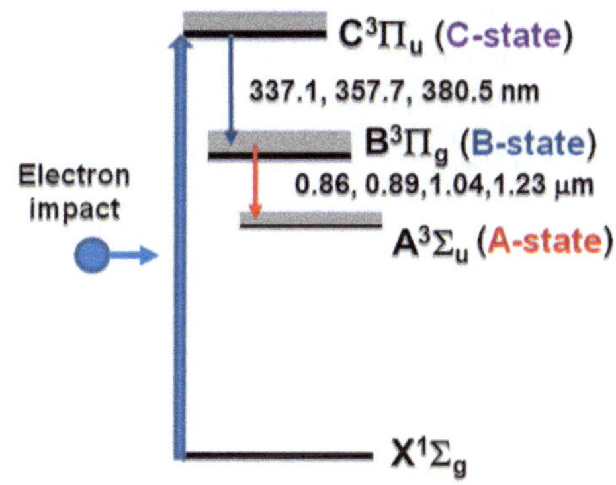

Fig. 5.2 System of electronic levels in N_2 molecule enabling laser generation

The ground $X^1\Sigma_g$ (X state) electronic state is a singlet; therefore, there is no possibility of a direct optical excitation to the upper triplet $C^3\Pi_u$ or $B^3\Pi_g$ laser levels. Thus, molecular nitrogen laser is ultimately a collisionally pumped gas laser. Typically, it operates as a discharge-type gas laser where the excitation and population inversion are achieved through an electron impact mechanism, as schematically shown in Fig. 5.2. However, more sophisticated collisional pumping mechanisms like electron beam pumping [9, 10], three-body recombination, and resonant excitation transfer can be realized (see below). A detailed description of the discharge-type nitrogen laser and the theory of its operation can be found in Ref. [11].

Concluding the Introduction, it should be mentioned that femtosecond filamentation is not the only possibility for standoff plasma generation and initiation of lasing in molecular nitrogen. One of the possibilities for a remote discharge is through the use of high-power microwave radiation [12]. However, to achieve an air breakdown at ambient pressure and at a long distance would require a microwave source of an unreasonably large size and power. Another possibility is to use optical pumping of lasing in the atomic products of dissociation of molecular nitrogen or oxygen [13–16]. Dissociation of nitrogen or oxygen molecules can be initiated by resonant two-photon absorption from a deep-UV laser source [13–15] or in a plasma of optical breakdown [16], whereas optical pumping of the atomic fragments is achieved by resonant two-photon absorption from the same deep-UV laser source. In principle, the predissociation step could be done in the plasma of a femtosecond filament. However, high absorption and scattering losses for the deep-UV pumping radiation in air set a practical limit on this otherwise very attractive scheme that yields an efficient bi-directional lasing.

5.2 Experimental Results on Lasing from Molecular Nitrogen in Femtosecond Filaments

The first indirect indications of lasing from molecular nitrogen initiated by femtosecond filaments in air were reported in [17]. In that experiment the filament was generated in air by a 42 fs laser pulses at 0.8 μm wavelength with pulse energy of

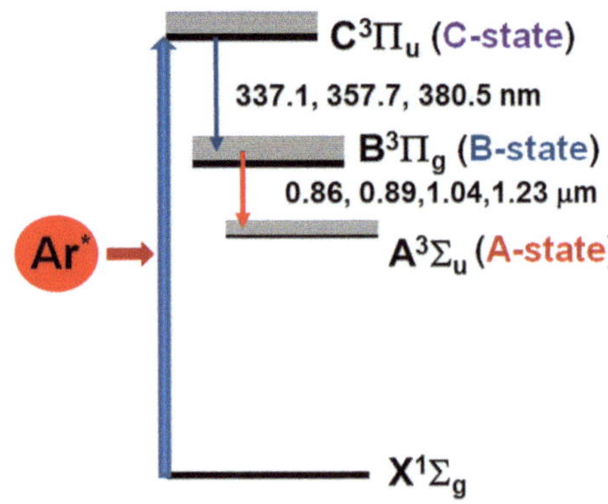

Fig. 5.3 The Bennett process of resonant excitation transfer in a collisionally pumped nitrogen laser

up to 25 mJ. The beam was focused by an $f = 1$ m lens, and the intensity of the backward-directed UV emission from the filament was measured with a photomultiplier (PMT), positioned behind a dichroic mirror, which was highly reflective for the near-IR laser radiation and highly transmitting in the UV spectral region. To select a particular emission band, different interference filters with 10 nm bandwidth were placed in front of the PMT. The conclusion about lasing from nitrogen in air was drawn from an exponential fit to the dependence of the fluorescence yield around 357 nm wavelength (corresponding to the $C^3\Pi_u(v = 0)$ to $B^3\Pi_g(v = 1)$ transition in N_2, where v notes the vibrational state) on the filament length. The filament length, in turn, was calculated from the laser power using Marburger's formula for the position of the nonlinear focus [18] $z_f \sim P^{-1/2}$ and the linear geometric optics equation for the focal length of the combined nonlinear and geometric focusing. The results and their interpretation raise many questions. The fact that the emission was so weak that it was only measurable by a photomultiplier does not speak in favor of it being due to laser generation. However, the publication [17] has spearheaded a new brunch of research in the physics of filamentation.

The first and so far the only well-characterized observation of backward-propagating stimulated emission initiated by a femtosecond filament in molecular nitrogen was reported in [19]. Preparing the plasma in a filament with the proper density and temperature and control over those parameters are the central problems for the realization of lasing, as it will be discussed in detail in the next section. In [19], the authors have circumvented these difficulties utilizing a well-known alternative scheme of pumping – the Bennett process of resonant excitation transfer [20].

The idea behind this approach is presented schematically in Fig. 5.3. Excited metastable atoms of Ar are produced as a result of a two-step kinetic process in the filament plasma involving three-body collisions $Ar^+ + 2Ar \rightarrow Ar_2^+ + Ar$ and dissociative recombination $Ar_2^+ + e \rightarrow Ar^*(4\,^3P_2) + Ar$. Population inversion in nitrogen is achieved through the reaction $Ar^*(4\,^3P_2) + N_2(X^1\Sigma_g) \rightarrow Ar + N_2^*(C^3\Pi_u)$. This process transfers the excitation energy of argon atoms to molecular nitrogen. Thus, excited argon atoms can provide a collisional pump for laser transitions in N_2 in a femtosecond filament, playing the same role as the role played by hot electrons in the discharge-pumped nitrogen laser. It is worth mentioning that the

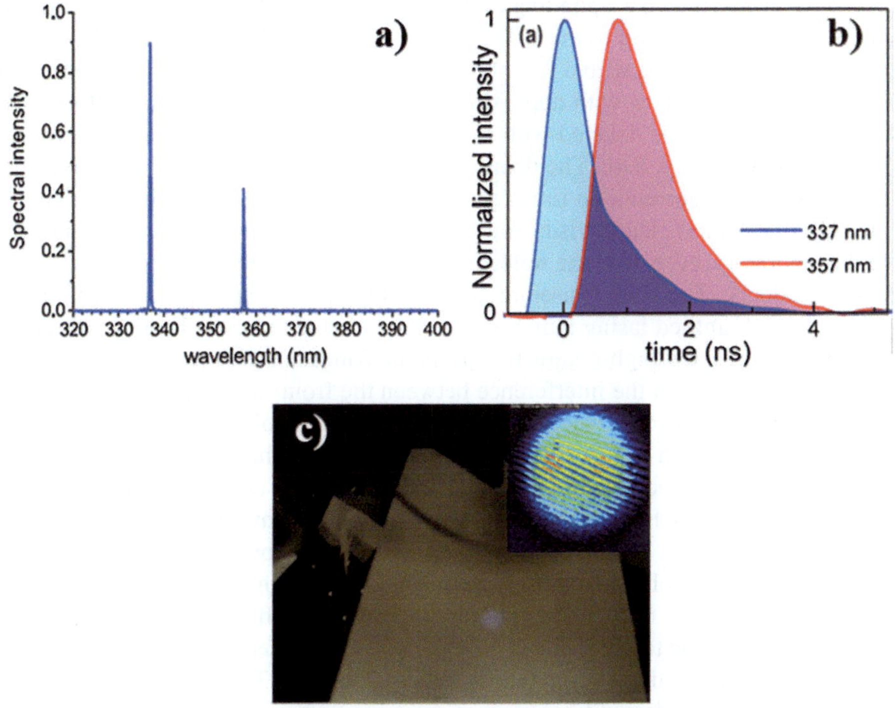

Fig. 5.4 (**a**) Spectrum of the bichromatic laser generation from nitrogen in the mid-IR filament. (**b**) Temporal profiles of the generated lasing pulses at the 337 nm and 357 nm wavelengths. (**c**) Photograph of the backward-directed UV lasing beam on a paper screen. The inset shows a CCD image of the beam. The fringe pattern in the beam is due to the interference between the two reflections from the front and back surfaces of the CaF_2 plate used to reflect the backward-propagating lasing emission

resonant excitation transfer mechanism of pumping is a very universal method used to achieve population inversion in gas mixtures. In particular, this is the pumping principle enabling the well-known He-Ne and CO_2 lasers.

Bi-directional lasing was demonstrated from filaments generated in a 4-m-long gas cell filled with a mixture of nitrogen under 1 bar partial pressure and argon under 5 bar partial pressure. Two distinctly different high-power femtosecond laser systems were used for filamentation, confirming the robustness of the lasing effect with respect to the wavelength of the filamentation source. One set of measurements was conducted with a near-infrared filament created by 200 fs, 6 mJ, 1.03 μm pulses from an Yb:CaF2 laser with 0.5 kHz repetition rate [21]. The second set of measurements was performed with a unique high-power mid-infrared (mid-IR) optical parametric chirped pulse amplification (OPCPA) laser system delivering 80 fs, 8 mJ pulses at a 20 Hz repetition rate at a wavelength of 3.9 μm [22]. The main results obtained with the mid-IR filaments are shown in Fig. 5.4. Backward-directed lasing from nitrogen was achieved for two wavelengths simultaneously, as shown in

Fig. 5.4a: at 337 nm corresponding to the $C^3\Pi_u(v = 0) \rightarrow B^3\Pi_g(v = 0)$ transition and at 357 nm corresponding to the $C^3\Pi_u(v = 0) \rightarrow B^3\Pi_g(v = 1)$ transition in N_2. Time-resolved measurements with a fast photodiode show that this emission is generated as a sequence of pulses with approximately 1 ns duration, where the pulse at the 357 nm wavelength is delayed with respect to the 337 nm pulse by approximately 1 ns, as shown in Fig. 5.4b. The delay can be explained by the lower gain and, correspondingly, longer buildup time for lasing on the 357 nm transition. The generated UV emission is clearly visible on a paper screen under the regular ambient light in the laboratory, even after the reflection from a CaF_2 plate placed into the beam at a 45° incidence. The plate is used to decouple the mid-IR filamentation beam from the backward-emitted lasing radiation. The measurements of the beam profile with a CCD camera show high-contrast interference fringes, as shown in Fig. 5.4c. The fringes originate from the interference between the front- and back-side reflections from the plate. These observations demonstrate high spatial coherence of the generated emission, which is one of the confirmations that what is observed is a laser generation and not spontaneous emission. The stimulated character of the emission is also confirmed by high-resolution spectral measurements showing narrowing of the emitted spectrum when spontaneous emission is suppressed by the stimulated emission process. Polarization measurements have shown that the polarization of the filament-initiated nitrogen laser is random, as expected for the single-pass free-space geometry. The total efficiency of the filament-initiated lasing in the backward direction was determined to be 0.5%, corresponding to 3.5 μJ of the total lasing energy in the two emission lines.

Numerical simulations based on the plasma kinetics in the filament show that under the experimental conditions, the chain of Bennett reactions requires approximately 1 ns buildup time to achieve population inversion, with a ratio of populations in the upper and lower lasing levels of about 10:1. The threshold for lasing is set by the interplay between the photon losses from the plasma channel and the gain. The latter is determined by the value of the population inversion, which enables nitrogen laser within the time window of approximately 4 ns. The subsequent drop in population inversion is mostly due to the quenching reactions $N_2(C^3\Pi_u) + N_2(X^1\Sigma_g) \rightarrow N_4 \rightarrow N_2(B^3\Pi_g$ vibrationally excited) $+ N_2(X^1\Sigma_g)$ and $N_2(C^3\Pi_u) + Ar \rightarrow N_2(B^3\Pi_g) + Ar$.

The Bennett pumping scheme can hardly be realized under real atmospheric conditions because it requires relative concentrations of nitrogen and argon that are far from the values naturally occurring in the atmosphere. Also, as it was shown in the above-mentioned experiments, the addition of more than 5% of oxygen terminates lasing through a very efficient (picosecond time scale [23]) quenching reaction $N_2(C^3\Pi_u) + O_2(X^3\Sigma_g) \rightarrow O + O + N_2(X^1\Sigma_g)$. Therefore, another pumping mechanism that would be efficient on the picosecond time scale, thus avoiding the negative influence of oxygen, should be found to enable atmospheric nitrogen lasing.

The realization of a direct analog of a discharge-type nitrogen laser in an optical filament has been reported in [24]. The experiment has been conducted using the COMET laser system, which is a part of the Jupiter Laser Facility at the Lawrence Livermore National Laboratory [25]. Laser pulses at 1.053 μm

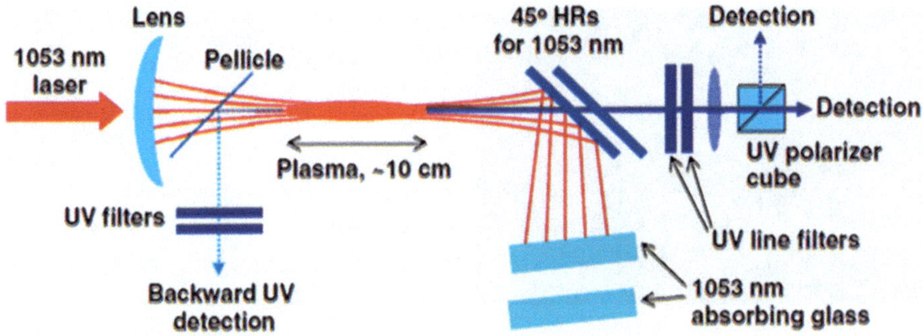

Fig. 5.5 Experimental setup used for the demonstration of lasing in neutral nitrogen under picosecond pumping

wavelength with a duration of 0.5 ps and energy of up to 10 J were stretched to 10 ps and focused directly in air by an $f = 2.6$ m lens. The picosecond pulse duration ensures collisional heating of the electrons in the filament plasma and, in addition to the strong-field ionization, avalanche ionization of air, similar to the situation in an RF discharge.

A sketch of the experimental setup is shown in Fig. 5.5. The spectral diagnostics of the UV emission from the filament in the backward direction were realized by using a pellicle beam splitter that had high transmission for the laser wavelength and about 10% reflectivity in the UV spectral region 340–400 nm. The backward-propagating emission, reflected from the pellicle, was additionally spectrally filtered with UV interference filters and focused onto the entrance slit of a UV spectrometer. The high-energy, forward-propagating fundamental radiation, as well as the white light continuum generated in the filament, was filtered by a pair of dichroic mirrors, highly reflective for the laser wavelength and highly transmitting in the 340–400 nm UV spectral region, and by two UV interference filters. The transmitted UV emission passed through a polarization cube and was focused onto the entrance slits of two UV spectrometers for simultaneous single-shot spectrum detection in the two orthogonal linear polarizations.

A single-shot side-view photograph of fluorescence from the plasma channel generated by a focused, 10 ps, 1 TW laser beam in air is shown in Fig. 5.6a. Bright, more than 10-cm-long plasma channel was observed on every shot. The transversal size of the plasma channel of 5 mm, inferred from single-shot burn patterns produced by the laser beam on an uncoated glass surface placed in the center of the Rayleigh zone of the beam, as well as the filament length were much larger than the vacuum spot size (about 40 μm) and the Rayleigh length of the beam (2.4 mm). A single-shot spectrum of the forward-propagating emission, polarized along the laser polarization, is shown in Fig. 5.6b. It shows a part of the broadband spectrum and the spectral tail of the third harmonic of the pump at 351 nm, transmitted by the interference filters and a clear 337 nm nitrogen laser emission line. The maximum energy measured in the 337 nm lasing emission was

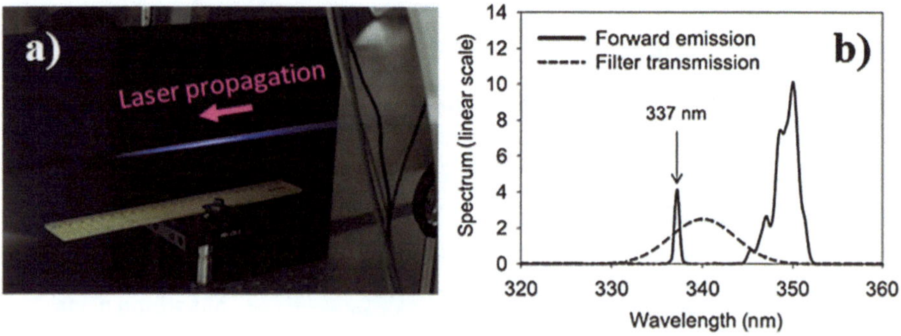

Fig. 5.6 (**a**) A single-shot photograph of plasma produced by a 10 ps, 1 TW laser pulse in air. (**b**) Measured single-shot spectrum of the UV emission from the picosecond filament. The dotted line shows the transmission curve of the interference filter

2.5 μJ. Observations at another possible emission line at 357 nm were obscured by the very intense third harmonic spectrum. No signal was detected in the orthogonal polarization, i.e., the observed lasing emission had the same polarization as the fundamental and the third harmonic radiation. Also, no signal was detected in the backward direction. These observations suggest that the laser generation from molecular nitrogen was seeded, in the forward direction, by the short wavelength tail of the broadband third harmonic radiation. This explains the observed polarization properties and suggests a suppression mechanism for the backward-propagating lasing. Another effect that would exclude backward-directed laser generation and that becomes especially important under the conditions of femtosecond filamentation is the traveling regime of excitation. This problem will be discussed in detail in the following sections.

Picosecond laser filaments enable standoff lasing in molecular nitrogen under the conditions closest to the conventional discharge-pumped laser. A very important difference, however, is that a picosecond laser filament is an example of an ultrafast discharge, when plasma kinetics during the igniter pulse is not influenced by the hydrodynamic process of expansion and heat transfer, both processes taking place on sub-nanosecond (or longer) time scale. Moreover, the picosecond electron heating time is significantly shorter than the time scale of the quenching reaction with oxygen (about 150 ps [23]), which is one of the most severe problems for achieving efficient lasing in air. That was confirmed in the Livermore experiment by the comparison of the lasing efficiency in air and in pure nitrogen under the same experimental conditions. Practical limitations for the application of picosecond filamentation to the problem of lasing in the atmosphere arise from the requirements to the filamentation source. While multi-terawatt level peak power (which is required for long-range filamentation in air) femtosecond laser systems are routinely commercially available, can operate at 10 Hz repetition rate, and can be compact and mobile (e.g., Teramobile project [26]), there are very few picosecond laser systems

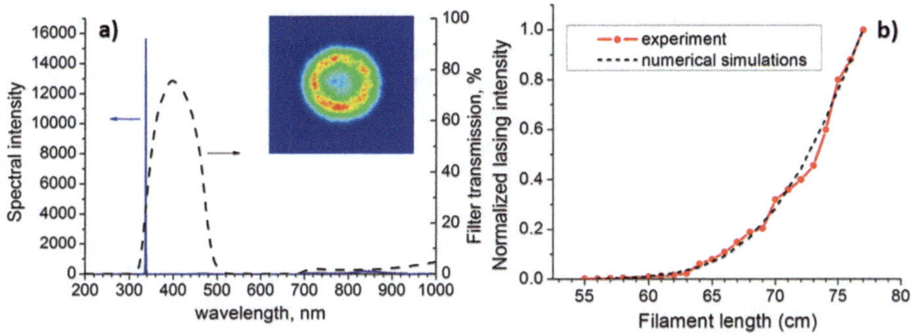

Fig. 5.7 (**a**) Spectrum of the UV lasing from pure nitrogen at 6.7 bar pressure (*blue line*) and spectral transmission of the interference filter (*dotted line*). The inset shows the CCD image of the far-field intensity profile of the 337 nm beam; (**b**) pressure dependence of the lasing intensity (*solid red line* with *circles*) and the result of numerical simulations (*dashed black line*).

of similar peak power. All of them are large-scale research facilities (like COMET) and operate under very low repetition rates (one shot in 5 min maximum repetition rate for the COMET system).

The first well-characterized forward-propagating lasing from femtosecond filaments in pure nitrogen pumped by electron collisions was reported in [27, 28]. The experiment was performed using the OPCPA mid-IR femtosecond system described above. The beam with a pulse energy of 8 mJ was focused by an $f = 100$ cm focusing lens into a 2-m-long cell filled with nitrogen. Forward-propagating lasing from molecular nitrogen at 337 nm wavelength was clearly observed for gas pressures above 5.5 bar, as shown in Fig. 5.7a. Note that filamentation at 3.9 μm wavelength for the given parameters of the laser pulse sets in at nitrogen pressure above 2 bar. To distinguish the lasing signal from the intense white-light continuum generated in the filament, a spectral filter transmitting in the 320–500 nm spectral range was placed in the beam after the gas cell. The lasing beam was clearly visible on a paper screen. Its far-field intensity profile, measured with a CCD camera, is shown in the inset in Fig. 5.7a. An increase of the gas pressure in the range from 2 bar to the maximum value of 7.7 bar, used in the experiments, results in a linear increase of the filament length. This was verified in a separate set of experiments, where the length of the plasma channels in the mid-IR filament as a function of the gas pressure was measured using the transversal capacitor probe technique [29]. The dependence of the lasing yield on the filament length is shown in Fig. 5.7b. The maximum energy of the 337 nm emission, measured with a calibrated UV photodiode at the highest gas pressure, was 70 nJ. Whereas lasing in the forward direction was measured for a broad range of gas pressures, no UV emission was detected in the backward direction.

Except for the case when the argon-assisted Bennett mechanism was used for pumping nitrogen lasing, relatively efficient lasing emissions from laser filaments in nitrogen, in the experiments described so far, were achieved using rather exotic laser sources – a high-power picosecond laser system and a long-wavelength femtosecond laser. However, the working horse of filamentation physics is a high-power, near-IR femtosecond system based on Ti:sapphire and operating at 0.8 μm wavelength. The quest for lasing in nitrogen under pumping through femtosecond filamentation at 0.8 μm wavelength was unsuccessful for a long time. The main focus in those investigations from the very beginning was on the backward lasing. A very important progress was achieved very recently, in 2014. First Mitryukovskiy et al. have demonstrated that the backward-emitted intensity at 337 nm increases dramatically when circular polarization is used for filamentation [30]. In that experiment, 50 fs pulses at 0.8 μm wavelength with energies between 2 and 9 mJ generated a filament in a gas cell filled with nitrogen or a nitrogen-oxygen mixture. The beam was focused by an $f = 100$ cm lens, and the spectrum of the UV emission from the filament in the backward direction as well as the transverse fluorescence from the plasma channel were measured with a monochromator equipped with a photo-multiplier (PMT) detector. It was shown that in the range of energies used in the experiments, the backward emission for a circularly polarized filamentation pulse grows much faster than that for a linearly polarized pulse, with the overall difference by a factor of 4 for the highest pumping energy. It was also confirmed that addition of oxygen lead to an abrupt drop of the emission efficiency for oxygen concentrations above 10%. The importance of circular polarization for efficient pumping of nitrogen emission was discussed in [31]. Experimental observations were supported by classical simulations of the energy distribution of the photoelectrons as a function of laser polarization. We will address this point in more detail in the next section where the physics of population inversion under pumping by laser filaments will be discussed.

Based on the results of [30], Ding et al. have confirmed the realization of population inversion in N_2 molecules in a femtosecond filament generated in a cell filled with nitrogen gas, through the measurements of amplification experienced by the backward-propagating seed UV pulse [32]. One of the very important results of that paper was a clear demonstration that the amplification and, therefore, the population inversion occur only for circular polarization in the filamentation laser pulse. In the experiments, 42 fs laser pulses at 0.8 μm wavelength with up to 12 mJ of energy generated a filament in a gas cell filled with nitrogen or air under atmospheric pressure. A small fraction of energy was used to generate the second harmonic pulse at 0.4 μm wavelength which was then additionally spectrally broadened via the nonlinear process of self-phase modulation in a glass plate, in order extend the spectrum to the 337 nm wavelength. A part of the spectrally broadened second harmonic was filtered by a 10 nm bandwidth interference filter with the transmission band centered at 337 nm, to form a seed pulse. The seed was focused into the gas cell in the direction opposite to the direction of propagation of the filamentation pulse. Amplification of the backward-propagating seed pulse by up to the factor of 200 was measured for the circularly polarized filament in pure nitrogen. No amplification was detected when the polarization of the filamentation pulse was linear. The authors have also measured the temporal dynamics of the gain by scanning the delay between the seed and the pump (filamentation) pulses. They have shown that the buildup time for population inversion in 1 bar nitrogen was about 4 ps, which is followed by a decay on the ~20 ps time scale. Population inversion was found to exist within a ~ 15 ps FWHM (full width at half maximum) time interval (or within ~40 ps foot-to-foot time interval).

Experiments in ambient air have demonstrated a forward-directed amplification, when the seed pulse was co-propagating with the filamentation pulse. However, in contrast to the measurements in pure nitrogen, no amplification for the backward-directed seed was observed, suggesting a significantly lower value of the gain in ambient air in comparison to that in pure nitrogen. Again, the negative influence of oxygen via collisional quenching was assumed to be the reason for poor amplification in air.

Simultaneously with [32], Yao et al. have published an experimental investigation of the gain dynamics in femtosecond filaments generated in pure nitrogen [33]. The experimental setup was similar to the one used in [32]. It was shown that population inversion can be reached only for close to circular polarization of the filamentation pump pulse. The quarter-wave plate that converted the initially linear polarization of the pump pulse into circular had to be aligned within ±10°, in order for the gain to be observed. The buildup time for population inversion was determined to be ~6 ps. After reaching a maximum, the gain decayed with the characteristic decay time constant of about 30 ps. As in the above-mentioned reports, addition of oxygen resulted in a dramatic shortening of the lifetime of population inversion and in the reduction of its maximum value.

Finally, forward-propagating lasing from femtosecond filaments in ambient air was demonstrated in experiments with the multi-TW femtosecond laser system JETI40 at the Friedrich-Schiller University Jena [34]. A multifilament bundle was generated by a 700 mJ, 30 fs laser beam at 0.8 μm wavelength in air. A photograph of the experimental setup is shown in Fig. 5.8. The 50 mm diameter laser beam was focused by an $f = 2.7$ m lens, placed in the vacuum chamber connected to the vacuum beamline of the laser system. The resulting plasma filament was about 50 cm long and had the diameter of several mm, estimated by the examination of a single-shot burn pattern on a thin glass plate. A quarter-wave plate placed after the output window from the vacuum beam line was used to change the polarization in the filamentation beam

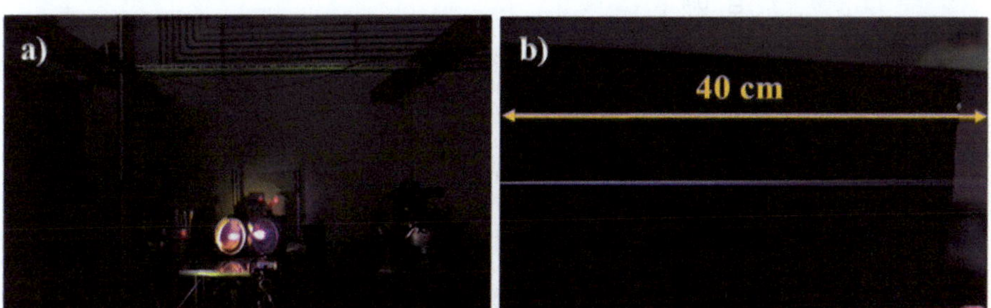

Fig. 5.8 (**a**) Photograph of the experimental setup used for the demonstration of nitrogen lasing in ambient air. (**b**) A zoomed-in photograph of the filament bundle

from linear to circular. The forward-propagating emission after the filaments was spectrally filtered by a pair of 45° dichroic mirrors highly reflective in the UV spectral range and highly transmitting in a broad spectral range around the fundamental 0.8 μm wavelength. The reflected radiation was focused by a lens into the entrance slit of a spectrometer covering the spectral range 270–1100 nm. For spectral measurements, an additional interference filter with 10 nm bandwidth at 337 nm wavelength or a band-pass filter transmitting UV radiation in the 320–385 nm spectral range was placed in front of the spectrometer. For the diagnostics of the backward-propagating lasing, a thin 45° dichroic mirror, with high transmission for the broadband fundamental spectrum and highly reflection for the UV, was placed in the laser beam before the quarter-wave plate.

The results of the measurements are presented in Fig. 5.9. The spectrum of the forward-propagating emission is shown in Fig. 5.9a. Filamentation is always accompanied by a dramatic spectral broadening and supercontinuum generation. In the

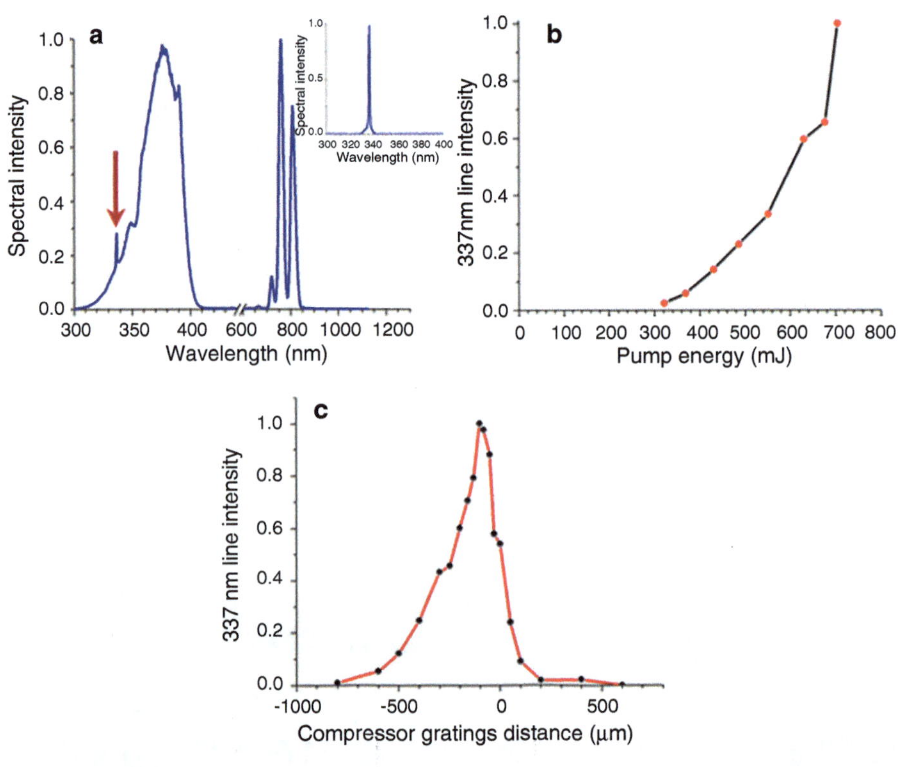

Fig. 5.9 (**a**) Single-shot spectrum of the forward-propagating emission from the filament after the dichroic mirrors and the spectral filter transmitting in the 320–385 nm spectral range. The *arrow* indicates the lasing line at 337 nm. The inset shows the lasing spectrum measured with a 10 nm bandwidth, 337 nm interference filter. (**b**) Dependence of the lasing intensity on the energy of the pump. (**c**) Dependence of the lasing intensity on the chirp of the filamentation pulse. Zero position of the gratings in the compressor corresponds to the shortest pulse. Negative values correspond to the larger and positive – to the smaller grating separations

case of multi-TW peak power, as shown in Fig. 5.9a, the continuum extends toward short wavelengths down to 300 nm. The lasing emission line at 337 nm was clearly resolved practically on every shot on top of the blue wing of the continuum spectrum, but only for circular or nearly circular polarization of the filamentation pulse. The change of the angle of the quarter-wave plate, which converted the polarization of the pump from linear to circular, of more than ±5° from the orientation for the perfect circular polarization, led to the complete loss of the lasing signal. The dependence of the forward-propagating lasing intensity on the pump energy is shown in Fig. 5.9b. A threshold for the single-shot lasing was observed at the pump energy of about 300 mJ, and the lasing signal growth was superlinear, increasing by an order of magnitude when the pump energy was changed from the threshold value of 350 mJ to the maximum value of 700 mJ used in the experiment. The maximum single-shot energy emitted at 337 nm was 70 nJ. Lasing signal was detected in the forward direction only. No measurable UV emission was observed in the backward direction. Similar to the experiments with picosecond filaments [24] discussed above, the failure to detect lasing in backward direction can be explained by the seeding of the lasing, in the forward direction, by the spectral continuum and by the traveling pump regime discussed in the next section.

To compare these results with the results of the experiments using a picosecond filament, the dependence of the lasing yield on the pulse chirp was investigated. The chirp was varied in both (positive and negative) directions from the zero value providing the shortest 30 fs pulse in the vacuum beam line. The longest duration of the stretched pulse, estimated from the third order autocorrelation measurements, was about 10 ps. As follows from the data shown in Fig. 5.9c, the highest lasing efficiency was reached when the distance between the compressor gratings is a bit longer than the value needed for a complete pulse compression, which corresponds to an initial negatively chirped laser pulse. This can be easily explained by the necessity to compensate the positive chirp introduced by the focusing lens, the output window from the vacuum beam line, and the dichroic mirror for the backward signal detection. In contrast to the Livermore experiment, the strongest lasing emission was observed for the shortest pulse generating the filament, and the efficiency dropped rapidly as the pulse chirp was increased. This difference can be explained by the difference, by an order of magnitude, in the filamentation pulse energy, assuming that a rather dense and thick plasma channel was required to generate sufficient gain for lasing.

Summarizing the review of the experimental results on lasing from molecular nitrogen in pure nitrogen gas or in air, pumped by femtosecond laser filaments, the following conclusions can be drawn:

- Population inversion is sensitive to the pumping wavelength. Relatively efficient lasing was observed for a long-wavelength mid-IR, linearly polarized filamentation pulses and was not observed for linearly polarized near-IR filamentation pulses having similar or even much higher peak power.
- Population inversion is sensitive to the polarization of the filament. More efficient lasing can be achieved with circular polarization. In the case of near-IR 0.8 μm

wavelength filamentation, circular polarization is the only possibility to achieve measurable stimulated emission even with multi-TW pump pulses.

- Similar to the case of conventional discharge-pumped nitrogen laser, oxygen influences the lasing dramatically, significantly reducing its efficiency at concentrations typical to those naturally occurring in the atmospheric air.

5.3 Physical Mechanisms Driving Dynamics of Population Inversion in N_2

Plasmas in femtosecond laser filaments are distinctly different from those produced in an ultrafast microwave discharge, in a pulsed high-voltage discharge or in the CW or nanosecond optical discharge. These differences stem from the properties of femtosecond laser filamentation in gases. First, due to the self-regulating balance between self-focusing and defocusing, the laser intensity in the filament is fixed (or clamped) at the level in the range 10^{13}–10^{14} W/cm^2 [4]. Second, due to the ultrashort (femtosecond) duration of the laser pulse generating the filament, the formation of plasma occurs through photoionization of the gas on the time scale of the pulse duration, i.e., long before the collisional electron heating and avalanche ionization kick in. The key physics involved in the buildup of the population inversion in nitrogen takes place in the afterglow of the ultrafast ionization. The plasma characteristics that govern population kinetics in the afterglow are entirely predetermined by the quantum-mechanical process of molecular ionization by the optical field.

The above features play the decisive role in the realization of standoff nitrogen lasing with femtosecond filaments. Electron collisions are now commonly accepted to be the main mechanism responsible for the buildup of population inversion in neutral nitrogen in femtosecond filaments [30–32]. The rate of collisional pumping depends on the temperature of the electrons. Since the laser intensity is clamped, the only two parameters that control the electron temperature in femtosecond filaments are the polarization and the wavelength of the driver laser pulse.

Before we proceed to the discussions of the dependence of the electron temperature on the details of the photoionization process, we mention another potential pumping channel based on the recombination of the cluster ion N_4^+. This recombination reaction, suggested first in [35], was considered previously as a candidate process that may enable nitrogen lasing in the filament plasma.

The production of the molecular cluster ion N_4^+ in the three-body collision process $N_2^+ + N_2 + N_2 \rightarrow N_4^+ + N_2$, with the reaction constant $\beta_4 = 5 \cdot 10^{-29}$ cm^6/s [36] is one of the most efficient processes in high-pressure nitrogen plasma. It was suggested that the dissociative recombination of N_4^+, $N_4^+ + e \rightarrow N_2(B,C) + N_2(X)$ leads to the production of excited molecular nitrogen predominantly in the C state and thus enables population inversion [35]. However, the proposition that the product of the recombination of N_4^+ has a high enough probability to end up in the excited state C is rather questionable. First, the variability of possible outcomes of the recombination reaction discussed in literature is quite large. For example, in [37] the result of the dissociative recombination is assumed to be in the ground state of nitrogen, i.e., $N_2(X)$ + $N_2(X)$ (see reaction 6 in Table IV in [37]). In [38], the result is $N_2(X)$ + $N_2(A)$ (reaction 9 in Table 1 in [38]). In [39], the result of the reaction is $N_2(X)$ + $N_2(A,B,C)$ with the same reaction constants for all three channels. There are also assumptions that the reaction products are $N_2(X)$ + N + N. The wide spread of possible scenarios signifies that the branching ratio is not known, and it is not clear whether it can be readily deduced from experiments. The cluster ion plays a decisive role in the process of electron recombination in nitrogen plasma, especially at high pressures, due to the very high value of the corresponding rate constant $\beta_{N_4^+} = 2{\cdot}10^{-6}\left(0.026/T_e\right)^{1/2}$ (cm^3/s), with the electron temperature in eV [36]. The recombination rate constant can be determined quite precisely, and this is the quantity of the primary interest in treating discharge problems in nitrogen plasmas. However, the question about branching of the products of dissociation remains open. On the other hand, if the assumption about the dominant production of the $N_2(C)$ is correct, a very effective lasing would be observed from almost any type of discharge in nitrogen, which contradicts the experimental evidence. Indeed, the process of formation of the N_4^+ cluster ion is one of the fastest and most efficient kinetic processes in nitrogen plasma. Therefore, the reaction of dissociative recombination, which is also very efficient for this cluster ion, would lead to population inversion regardless of the electron temperature and density. This conclusion is against experimental evidence as the discharge-pumped nitrogen laser, for both low and high nitrogen pressures, is confirmed to be enabled by the electron impact excitation mechanism (cf. [11]). It is worth to mention that the role of the dissociative recombination of the N_4^+ cluster ion diminishes under the conditions of atmospheric filamentation, where the majority of the plasma is due to ionization of oxygen molecules. All of the above arguments lead to the conclusion that population inversion through dissociative recombination of the N_4^+ cluster ion cannot be efficiently realized in ultrafast optical filamentation.

Going back to the discussion of the pumping mechanism involving electron collisions, we recall that in a femtosecond filament, the initial electron energy distribution function (EEDF) of the photoelectrons is a solution of the quantum-mechanical problem of the ionization of a gas by a high-intensity ultrashort laser pulse. This solution fully determines the influence of the wavelength and polarization of the driver laser on the prospects of reaching population inversion and on the entire initial plasma dynamics in the filament. The full 3D numerical simulations of this multi-electron, multi-nuclei problem are extremely challenging. We are not aware of published reports on such numerical studies for molecular gases. From the theory of strong field ionization [40, 41], it is known that the ionization probability and the momentum/energy distribution of the photoelectrons are essentially governed by two parameters: the Keldysh parameter $\gamma = \omega\sqrt{2m_e W_i}/eE = \sqrt{W_i/2U_p}$ (where ω and $E(t)$ are the frequency and electric filed of the laser pulse, respectively, W_i is the ionization potential, and U_p is the ponderomotive energy) and polarization of the laser field. A powerful analytical approach has been developed in [42] that allows one to compute the energy distribution function of the photoelectrons for arbitrary values of the Keldysh parameter γ and arbitrary polarization of the laser electric field for atomic gases. This approach can be used, as a good approximation, for molecular gases as well, because in the ensemble of randomly oriented molecules colliding with electrons, subtle features of the EEDF related to the specific molecular structure are not expected to play an appreciable role.

5.3.1 Polarization Dependence of EEDF

The general role of polarization of the ultrashort driver pulse in the energy distribution of photoelectrons has been first pointed out in [43], and in the specific application to the filament-initiated lasing in nitrogen – in [31]. A calculated EEDF of photoelectrons in the plasma filament produced by a 0.8 μm femtosecond laser pulse with intensity of $7 \cdot 10^{13}$ W/cm^2 in nitrogen is shown in Fig. 5.10, for different polarization ellipticities ξ of the laser field [42]. Also shown are the experimentally measured electron impact excitation cross-sections for the B and C electronic states of the nitrogen molecule [44]. In the case of linear polarization, the temperature of the photoelectrons is low, and as it will be shown below by numerical simulations, no population inversion can be achieved if a near-IR pump is used. However, when an elliptically polarized laser pulse is used to generate the filament, the EEDF can have a significant content of high-temperature electrons with the maximum energy $\varepsilon_{max} \approx 2U_p\xi^2$ [42]. As a result, a significant fraction of photoelectrons has kinetic energy in the range 11–40 eV, which is the range that is favorable for creating population inversion in neutral nitrogen molecules.

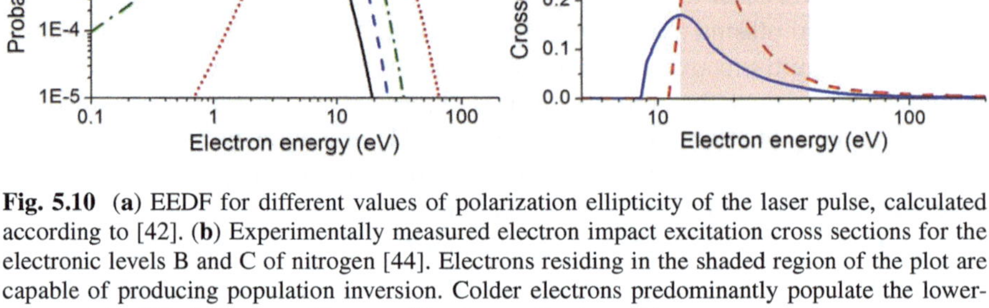

Fig. 5.10 (**a**) EEDF for different values of polarization ellipticity of the laser pulse, calculated according to [42]. (**b**) Experimentally measured electron impact excitation cross sections for the electronic levels B and C of nitrogen [44]. Electrons residing in the shaded region of the plot are capable of producing population inversion. Colder electrons predominantly populate the lower-energy state of the lasing transition, resulting in negative inversion

5.3.2 *Dependence of the EEDF on the Wavelength of the Driver*

The second parameter of the laser pulse that directly affects the temperature of the photoelectrons is the laser wavelength. In the tunneling ionization limit ($\gamma \ll 1$) and for a linearly polarized driver, the EEDF of the photoelectrons takes the following approximate analytical form: $f_0(\varepsilon) \propto \exp\left\{-\frac{2}{3}\gamma^3 \frac{\varepsilon}{\hbar\omega}\right\}$ [42]. Therefore, the initial effective electronic temperature $T_e \propto \hbar\omega/\gamma^3 \propto \lambda^2$ is implying that a longer-wavelength filament contains hotter plasma, as shown in Fig. 5.11. The most prominent feature of the EEDF dependence on the wavelength of the driver, λ_L, shown in Fig. 5.11, is the orders of magnitude higher concentration of electrons with energies above 11 eV, which are necessary to pump population inversion in nitrogen, in mid-IR filaments compared to the case of the near-IR around $\lambda_L = 0.8$ μm.

Thus, by choosing the wavelength and polarization of the laser pulse generating the femtosecond filament, it is possible to reach plasma temperature that is sufficient for efficient collisional pumping of population inversion in neutral nitrogen. The temporal dynamics of the nitrogen lasing and the realization of bi-directional or unidirectional generation are directly related to the plasma evolution in the afterglow. Since plasma kinetics proceed in the afterglow, under field-free conditions, the time scale of the relaxation of the electron temperature is determined by different nonelastic collision processes and recombination. Measurements using optical and THz interferometric methods have shown that a filament-generated plasma in air or pure nitrogen decays on the time scale of few hundred picoseconds [45, 46]. As reported in [32, 33], the gain lifetime is several tens of picoseconds. These observations highlight the traveling-wave regime of electronic excitation of nitrogen, which makes the backward-propagating lasing very hard to achieve. The traveling-wave pumping is a general feature of laser systems with fast decaying gain. It is the

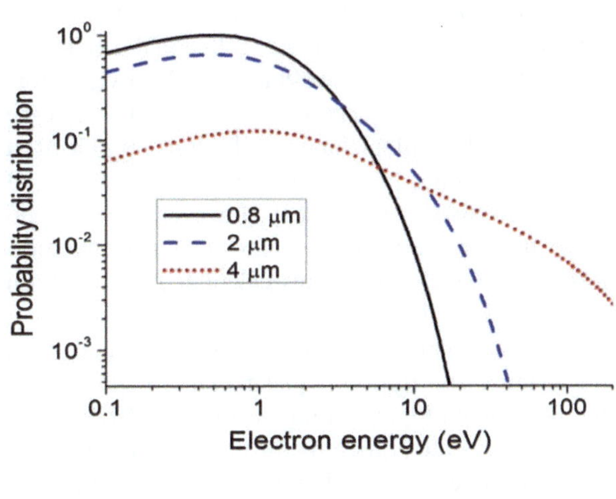

Fig. 5.11 EEDF for different wavelengths of a linearly polarized driver pulse, calculated according to [42]

major limitation in single-pass X-ray lasers based on laser-generated plasmas. The photon, spontaneously emitted at the proper time at the beginning of the filament and co-propagating with the pump pulse, will experience maximum population inversion along the entire length of the filament, resulting in efficient forward-propagating stimulated emission. At the same time, the photon spontaneously emitted in the backward direction will interact with the rapidly decaying population inversion. The effective gain length for such a photon will be only few centimeters, corresponding to the picosecond scale gain lifetime. The traveling-wave pump regime has been also experimentally demonstrated for pumping by a pulsed microwave discharge, in which case at least one order of magnitude difference between the gains for the forward-propagating and backward-propagating stimulated emissions has been reported [12].

5.4 Theory of the Filament-Initiated Neutral Molecular Nitrogen Lasing

A theoretical model describing the operation of the atmospheric nitrogen laser has been first suggested in [47, 48]. The model was based on the coupled system of rate equations for the electron density, the density of the thermal energy of the electrons, and the densities of excited N_2 molecules and ionic nitrogen species. The model accounted for the heating of the electrons by an additional heater laser pulse, in order to treat the igniter-heater concept, which we will discuss in the next section. The evolution of the generated UV emission was described by the radiation transfer equation coupled to the rate equations by the spontaneous and stimulated emission rates. The crucial assumption used in the calculation of the collision rates was the Maxwellian EEDF. It is known from discharge physics that in molecular gases, where inelastic collisions play the dominant role in the energy exchange between electrons and molecules, the EEDF is not Maxwellian [49]. The deviation from Maxwellian EEDF is primarily in the distribution of high-energy electrons, which play the decisive role in the collisional excitation of nitrogen molecules. The assumption of the Maxwellian EEDF provides for good quantitative agreement between theory and experiment for the discharge-pumped nitrogen laser [11, 50] because of the replenishment of high-energy electrons through joule heating in the discharge electric field. A similar situation should take place in the case of optical heating considered in [47, 48]. However, the assumption of Maxwellian EEDF fails under the conditions of the field-free plasma evolution in the afterglow of the femtosecond filament. The rates calculated based on this assumption will differ dramatically from the correct values.

Theoretical analysis of the filament-initiated single-pass nitrogen laser based on the ab initio description of plasma kinetics has been proposed in [27]. In the model, the Boltzmann kinetic equation for the EEDF is self-consistently solved together with a set of rate equations for the populations of different electronic states in the nitrogen molecule, and for the densities of nitrogen and oxygen ions and atoms.

The analytical solution from [42], discussed above, is used as initial condition in the kinetic equation for the EEDF. A set of single-pass laser equations is coupled to the kinetic equations in order to calculate the characteristics of the stimulated and spontaneous emissions from the plasma and to compute the stimulated emission rates used in the rate equations. The ultrashort time scale of plasma formation in the femtosecond filament allows for the simplification of the kinetic equation through the use of the Lorentz approximation, which reduces the problem to the solution of a one-dimensional kinetic equation for the EEDF [49].

The kinetic equation for the EEDF describes the time-dependent balance of the electron fluxes along the energy axis, accounting for the elastic collisional cooling, electron-electron Coulomb collisions, and inelastic collisional cooling. The inelastic collision processes included in the model are the electronic excitations from the ground electronic state to the $A^3\Sigma_u$, $B^3\Pi_g$, $a^1\Sigma_u$, $a^1\Pi_g$, $C^3\Pi_u$ electronic states of molecular nitrogen, super-elastic scattering from those electronic states, vibrational excitation of nitrogen and oxygen molecules, electron impact ionization, and electron-ion recombination.

The rate equations describe the balance of populations for the electronic states in N_2, listed above, and for different ionic and atomic species in the filament plasma, namely, N_2^+, O_2^+, N_4^+, and atomic nitrogen and oxygen. A relatively short plasma and gain lifetimes in femtosecond filaments make it feasible to select only few dominant kinetic processes that contribute to the population dynamics of various electronic states in molecular nitrogen on the time scale of several nanoseconds. The following kinetic processes are included in the model:

1. Collisional quenching of the C and B states (the upper and the lower lasing levels in the nitrogen laser) through the collisions with nitrogen molecules in their ground state: $N_2(C^3\Pi_u) + N_2(X^1\Sigma_g) \rightarrow N_2(a^1\Sigma_u) + N_2(X^1\Sigma_g)$
2. Energy pooling reactions $N_2(A^3\Sigma_u) + N_2(A^3\Sigma_u) \rightarrow N_2(B^3\Pi_g, C^3\Pi_u) + N_2(X^1\Sigma_g)$
3. Production of the molecular cluster ion N_4^+ through the three-body collision process $N_2^+ + N_2 + N_2 \rightarrow N_4^+ + N_2$ and its dissociative recombination $N_4^+ + e \rightarrow N_2(B,C) + N_2(X)$
4. Dissociative recombination of an electron with a nitrogen molecular ion through two reaction channels: $N_2^+ + e \rightarrow N + N$, providing dissociation products in the ground electronic state of the atoms, and $N_2^+ + e \rightarrow N + N(^2D)$
5. Recombination channel involving the cluster nitrogen ion N_4^+ through the following reaction: $N_4^+ + e \rightarrow N_2(X) + N_2(B,C)$ with equal branching for the excited molecular products in the B and C electronic states

The system of single-pass laser equations is the system of one-dimensional (along the filament) radiation transfer equations describing the evolution of the backward- and forward-propagating stimulated emissions.

The results of numerical simulations for the femtosecond filament, generated by 0.8 μm laser pulses with different polarizations in pure nitrogen under 1 bar pressure, are shown in Fig. 5.12 [27]. Population inversion can be reached for laser ellipticities greater than 0.5. Both the peak value and lifetime of inversion increase when polarization approaches the circular state, as shown in Fig. 5.12a. For nearly

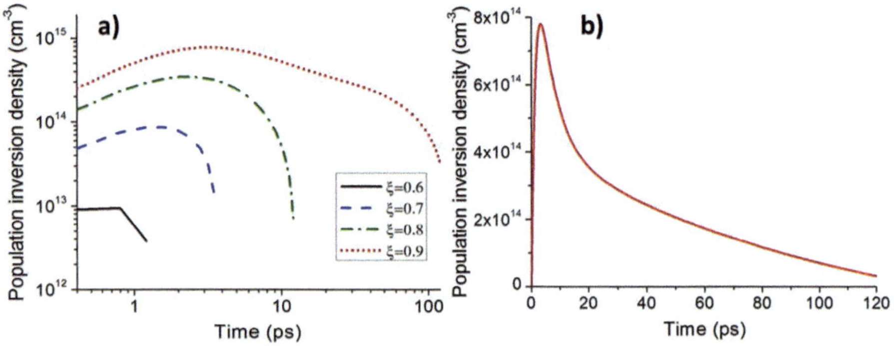

Fig. 5.12 (**a**) Time dependence of the population inversion density for different polarization ellipticities ξ of the filament pulse. (**b**) Population inversion density for ξ = 0.9. The calculations are done without accounting for the contribution from stimulated emission

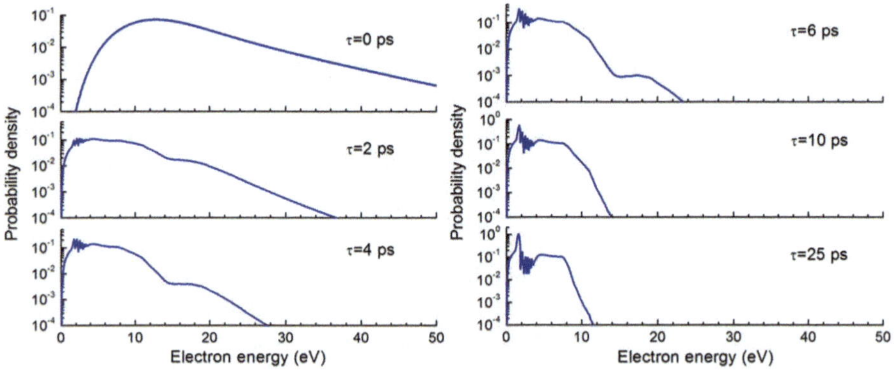

Fig. 5.13 Time-dependent evolution of the EEDF. The oscillatory structure in the 2–4 eV energy range resembles resonances for vibrational excitation of molecular nitrogen

circular polarization, the population inversion density reaches a maximum approximately 8 ps after the pump pulse, then drops by one half within 16 ps and slowly decays during the next 100 ps. This behavior of population inversion is in good quantitative agreement with the results of time-resolved gain measurements reported in [32, 33], which we discussed above. It can be understood by analyzing the dynamics of the EEDF (Fig. 5.13). The EEDF formed by photoionization has an initial temperature of 10 eV with more than 50% of electrons having initial energies in the 11–40 eV range, which is favorable for collisionally pumping population inversion. Within the initial 6 ps, the hot electrons from the high-energy tail of the EEDF vanish due to the loss into impact excitation of the electronic excited states of molecular nitrogen. Electrons with energies between 1.5 eV and 5 eV are efficiently consumed through losing their energy into vibrational excitations of nitrogen molecules, which is evidenced by the appearance of an oscillating resonant structure in the EEDF over the corresponding energy interval (Fig. 5.12). Since the excitation of the upper lasing state C consumes the largest electron kinetic energy, the amount of electrons contributing to populating the C state becomes much smaller than the amount of electrons contributing to populating the lower lasing state B. This leads to a fast drop of population inversion within next several tenses of picoseconds. Then, when the energetic tail of the EEDF loses electrons with energies above 8.5 eV, the decrease in population inversion slows down because it is caused now by slower processes of (i) electron impact de-excitation and (ii) collisional quenching. The latter leads to the general reduction of populations of both B and C states. In general, the fast cooling of electrons due to the electronic and vibrational excitations splits the EEDF into two distinct parts: a prominent peak of cold electrons with energies about 1 eV, which is insufficient for any excitation by an individual collision, and a group of relatively hot electrons with energies between 4 eV and 10 eV. In the latter region, the electron energy is too high for vibrational excitation. At the same time, the energy outflow due to the electronic excitations of the A and B states is nearly balanced out by the electron energy inflow through collisional de-excitation. The EEDF continues to cool down, without significant further reshaping, on a much longer time scale determined by elastic collisions. It is worth mentioning that this EEDF is significantly non-Maxwellian.

The development of the forward-propagating emission along the filament is shown in Fig. 5.14. Stimulated emission dominates over spontaneous emission for the filament length about 1 cm. Stimulated amplification proceeds in the linear regime (i.e., grows exponentially with the length of the gain region) for the filament length of up to 10 cm. The small signal gain extracted from the simulations is about 2.2 cm^{-1}. This small signal gain value is in good agreement with the results of the measurements of backward-seeded amplification reported in [32]. It would enable amplification by the factor of 200 for an amplification length of 2.4 cm, which is defined by the population inversion lifetime of 80 ps, in agreement with the values from [32]. For a longer filament, a saturation regime of forward-propagating lasing will be reached. At the same time, only low-intensity spontaneous emission is propagating in the backward direction.

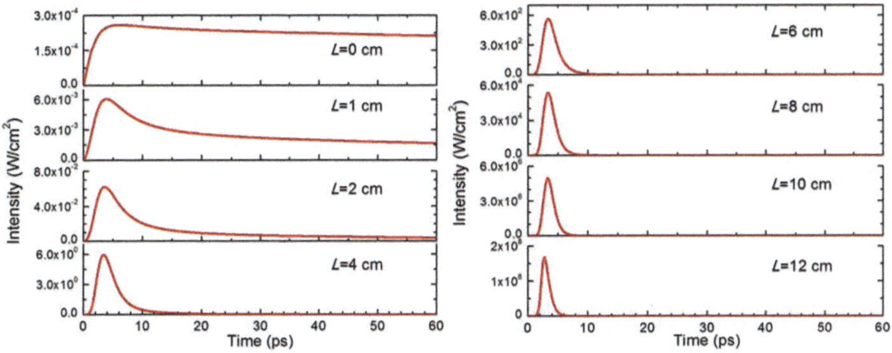

Fig. 5.14 Development of the stimulated emission pulse along the propagation length L in the filament plasma for the case of 800 nm pumping

In addition to the case of pumping at 0.8 μm wavelength, discussed above, the model adequately reproduces the experimental results on pumping nitrogen lasing with a mid-infrared laser source shown in Fig. 5.7. The computed efficiency of lasing is lower than that in the case of pumping with a circularly polarized 800 nm filament. We argue that due to the λ^2-scaling of the plasma refraction, plasma density in the mid-IR filament is expected to be significantly lower than that in the near-IR case, with the corresponding reduction of the lasing efficiency. The results of numerical simulations for plasma kinetics in a linearly polarized 4 μm filament are shown in Fig. 5.15. Low plasma density greatly reduces the efficiency of the direct excitation by electron impact, as shown in Fig. 5.15a. An alternative pumping channel, which is operative on much longer time scales, becomes important. This mechanism is through the energy pooling reaction (reaction 2 in the list of kinetic processes above). According to [51], that reaction has a higher rate for the production of nitrogen in the C state than in the B state. In addition, the collisional quenching rate for the B state is higher than that for the C state [36]. (Note that although this reaction is present in the near-IR filament, its role in pumping the lasing is negligible compared to the direct channel due to the much higher plasma density in that case.) The energy pooling mechanism has not been reported for the case of discharge-pumped nitrogen lasers and appears to be a unique feature of ultrafast optical discharge in femtosecond filaments. Nearly equal contributions to the buildup of population inversion from the two mechanisms lead to the formation of a two-pulse sequence in stimulated emission. The first shorter pulse originates from the direct electron impact excitation, while the second, much longer pulse is generated due to the plasma-chemical dynamics, as shown in Fig. 5.15b. However, the total lasing efficiency is rather low and the peak intensity of the stimulated emission pulse is only several tens of kW/cm^2 even for a 1-m-long filament, primarily because of low plasma density. In contrast to lasing from the filament plasma generated by an elliptically polarized near-IR femtosecond laser pulse, where the pumping mechanism is similar to pumping by an energetic electron beam, pumping by linearly polarized mid-IR filament is similar to the electrical discharge pumping.

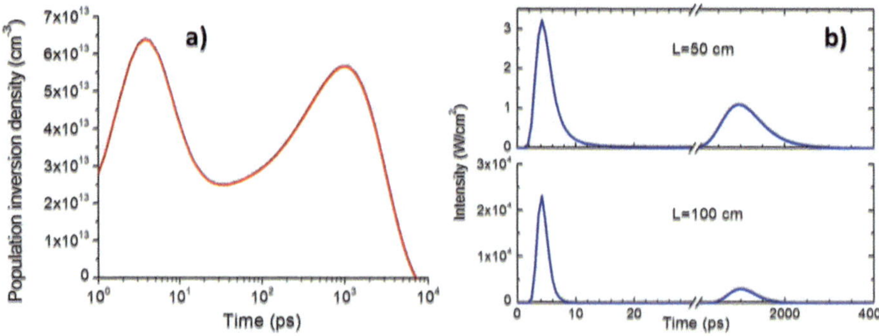

Fig. 5.15 (**a**) Time-dependent population inversion density and (**b**) time-dependent intensity in the stimulated emission pulse in nitrogen at 1 bar pressure, pumped by a 4 μm filament

The above model can be straightforwardly generalized to include more plasma-chemical reactions. In particular, the addition of reactions related to oxygen allows simulations of lasing in air filaments. There are several factors that reduce the efficiency of electronic excitation of nitrogen dramatically when oxygen is added. In air, ionization of oxygen is the main source of plasma. Due to the lower ionization potential of oxygen, the clamping intensity in the air filament is lower than that in the filament in pure nitrogen, leading to colder EEDF. Also, the density of cluster ions N_4^+ is rather low in the air filament. Correspondingly, their contribution to the production of excited nitrogen molecules in the B and C states becomes negligible. Another major effect in the air filament is quenching of population of the lasing levels by collisions with oxygen molecules. Both the upper and the lower lasing levels are emptied with an approximately equal efficiency through the reaction $N_2(C,B) + O_2 \rightarrow O + O + N_2(X^1\Sigma_g)$ [36]. The time constant for quenching by collisions with oxygen molecules has been measured to be about 150 ps [23], which is significantly longer than the time scale of electron pumping. Therefore, quenching by collisions with oxygen cannot appreciably affect the electron pumping channel, although it can arrest the energy pooling mechanism that we suggested for the case of mid-IR pumping. It has been experimentally found that the addition of several percent of oxygen to the nitrogen pumped by a circularly polarized 800 nm filament results in a dramatic suppression of nitrogen lasing [32, 33]. A detailed investigation of the filament dynamics in a nitrogen-oxygen gas mixture at various partial gas pressures is required in order to explain these findings.

5.5 Nitrogen Lasing in Laser Filaments Heated by Microwaves

As follows from the measurements of the gain dynamics and from the theory of the filament-initiated lasing discussed above, the short plasma lifetime in the filament is the main stumbling block toward the realization of efficient backward-propagating lasing in molecular nitrogen. One possible way to circumvent this problem could be through the use of the recently proposed "igniter-heater" concept [47, 48]. This concept, which is well known in the laser fusion community, involves the generation of plasma by a femtosecond laser pulse (the igniter) and the subsequent heating of the plasma by an additional radiation source (the heater). The heater pulse extends the plasma lifetime through collisional heating of the electrons, thus preventing recombination and plasma-chemical processes that lead to the plasma decay. The experimental realization of this idea in a He-N_2 gas mixture at the atmospheric pressure has been reported in 1974 [52]. In that experiment, the plasma was generated by an electrical discharge and subsequently collisionally heated by a high-power CO_2 laser, resulting in population inversion and lasing in the second positive system of molecular nitrogen. Differently from the cases of a DC or RF discharge, where plasma density is low, in femtosecond filaments typical plasma densities are $\geq 10^{16}$ cm^{-3}, and optical plasma heating faces fundamental difficulties related to the plasma refraction of the heater beam. Optical heating can efficiently increase the temperature of a neutral gas in a filament-assisted channeling of a high-voltage DC discharge [53] and significantly extend the lifetime of filament plasma [54], but no successful application of optical heating to the problem of lasing in neutral molecular nitrogen has been reported to date.

An alternative approach to plasma heating in femtosecond filaments in air that utilizes a microwave discharge has been proposed in [55]. The idea was inspired by the experimental realization of nitrogen lasing in air under pumping by high-power microwaves inside a microwave waveguide [12]. The numerical treatment of this approach, which we discuss here, is based on the solution of the kinetic equation for the EEDF. In essence, it is an extension of the formalism from [27] discussed above, with the addition of the effect of heating by a microwave field and of the rate equations describing the processes related to the presence of oxygen in the air. In addition, the set of radiation transfer equations is replaced by the rate equation for the number of emitted UV photons, which is similar to the treatment discussed in [34, 56].

The results of numerical simulations show that free-space lasing from molecular nitrogen in air enabled by microwave-heated plasma in a femtosecond filament is indeed possible but only if several conditions are satisfied. First, typical plasma density in the femtosecond air filaments generated by near-IR laser pulses is too high for the buildup of population inversion to occur. Plasma density in filaments is rather sensitive to the focusing conditions [57]. For near-IR laser sources, it ranges from ~10^{17} cm^{-3} for the case of relatively tight focusing (which is typical for small-scale laboratory experiments) down to ~10^{15} cm^{-3} for the case of self-focusing of a collimated or almost collimated laser beam (which is the situation typical in open-field experiments with TW-level peak power laser systems like TERAMOBILE). Numerical simulations show that for plasma densities above 10^{15} cm^3 population, inversion is strongly suppressed or completely quenched by electron collisions. Remarkably, the same electron impact mechanism is responsible for both pumping population inversion at relatively low plasma densities and for terminating the inversion at high plasma densities. This effect and the threshold in plasma density ~10^{15} cm^3 are well known in the field of discharge-pumped nitrogen lasers [58]. Therefore, time delay between the femtosecond laser pulse that generates plasma through filamentation and the microwave heater pulse is necessary to reduce plasma density to the value below 10^{15} cm^{-3}. The value of the time delay is determined by the plasma decay rate, which, in turn, depends on the initial plasma density. For the initial plasma density in the filament of 10^{17} cm^3, the decay to the 10^{15} cm^3 level takes about 2 ns [45], whereas for softer focusing conditions and the initial plasma density of 10^{16} cm^{-3} it will take more than 10 ns.

The temperature and distribution dynamics of the free-evolving filament plasma with the initial plasma density of 10^{16} cm^3 are shown in Fig. 5.16. These dynamics occur before the arrival of the microwave heater pulse. The filament is generated by a linearly polarized femtosecond laser pulse at 0.8 μm center wavelength, with the value of clumped intensity set to $7 \cdot 10^{13}$ W/cm^2. As evident in the Fig. 5.16, the aver-

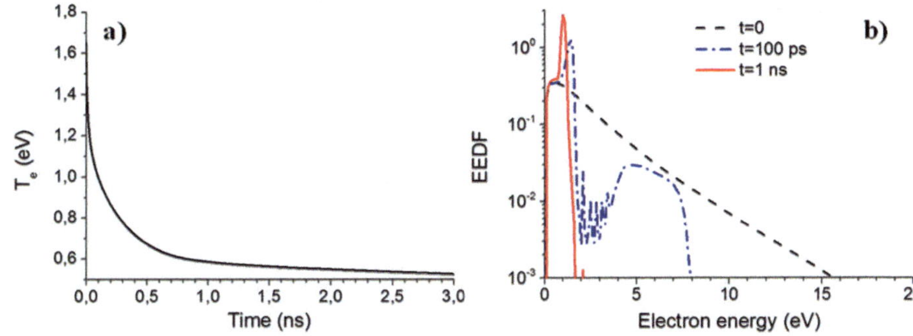

Fig. 5.16 (**a**) Time dependence of the electron temperature in the free-evolving filament plasma. (**b**) Time dependence of the EEDF. The analytical solution from [42] is taken as the initial distribution function at $t = 0$. The apparent oscillatory structure at $t = 100$ ps is due to the resonant vibrational excitations of molecular nitrogen

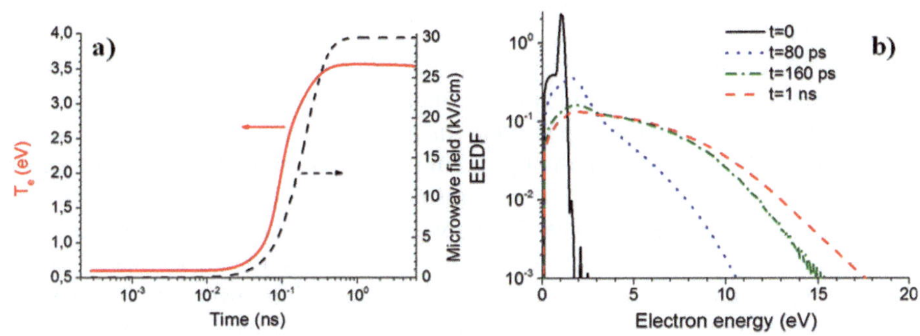

Fig. 5.17 (**a**) Time evolution of the electron temperature in the filament plasma heated by 30 GHz microwaves. The *black dashed line* shows the temporal profile of the microwave pulse. (**b**) Corresponding time evolution of the EEDF

age electron temperature stabilizes at the value of $\approx$0.6 eV within approximately 1 ns, at which point the EEDF becomes quasi-stationary.

The evolution of this initially cold EEDF due to the heating by 30 GHz microwave radiation with an electric field amplitude of 30 kV/cm (which corresponds to the breakdown threshold for the atmospheric pressure air) is shown in Fig. 5.17. Here we assumed that the microwave heater pulse arrives with the delay of several nanoseconds after the femtosecond filamentation pulse, at which point the plasma density is reduced down to $2.5 \cdot 10^{14}$ cm^{-3}. Heating the plasma to the stationary electron temperature of $\approx$3.5 eV takes about 1 ns. The resulting hot and stationary EEDF is rather insensitive to the parameters of the initial cold distribution because of the diffusive nature of the kinetic equation.

Another important parameter that, in addition to plasma density, dramatically affects nitrogen lasing from a microwave-heated laser filament is the amplitude of the microwave field. As an example, a 16% reduction in the field

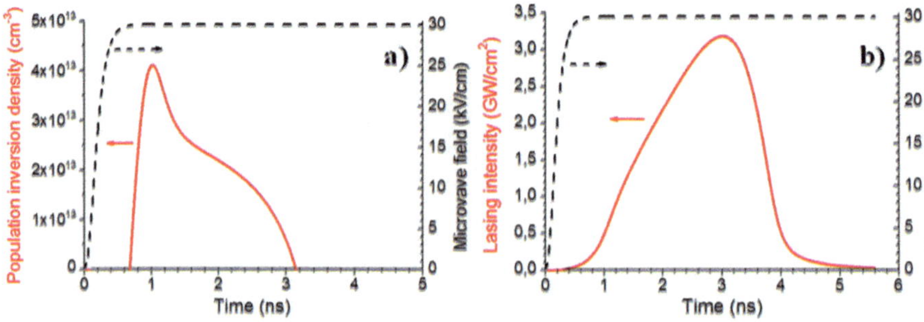

Fig. 5.18 Development of lasing with initial plasma density of $2.5 \cdot 10^{14}$ cm^{-3}. (**a**) Temporal evolution of population inversion density (*solid line*) and (**b**) the lasing pulse at 337 nm wavelength (*solid line*). The dashed lines show the temporal profile of the microwave pulse.

amplitude from its breakdown value (i.e., the reduction from 30 kV/cm to 25 kV/cm) makes the population inversion in the atmospheric nitrogen unattainable.

Finally, the finite length of the plasma filament sets the minimum value of gain that is necessary for the development of stimulated emission within the time interval which takes a spontaneous photon to traverse the filament. As numerical simulations show, the filament length of ≥10 cm is required in order for the stimulated emission to prevail over fluorescence, resulting in the laser-like emission. The time evolution of population inversion between the C and B electronic states in N_2 is shown in Fig. 5.18a, for the initial electron density of $2.5 \cdot 10^{14}$ cm^{-3} and a 10-cm-long filament. The buildup time for population inversion is ~0.3 ns. Inversion then rapidly declines within approximately 0.7 ns, followed by a slow decay within the next ~2.5 ns. The fast decay manifests the onset of saturation of amplification by stimulated emission, while the slow decay results from various collisional quenching processes, electron impact ionization and deactivation. The small signal gain, calculated for the peak value of population inversion of $4 \cdot 10^{13}$ cm^{-3}, is 0.11 cm^{-1}. This value is in good agreement with the measured gain in a discharge-pumped nitrogen laser operating under similar plasma densities [59]. As shown in Fig. 5.18b, our simulation predicts the generation of a 337 nm laser pulse with the duration of several nanoseconds and the total energy of about 1 mJ, assuming 100 μm filament diameter.

The suppression of lasing at high plasma density, discussed above, is demonstrated in Fig. 5.19. In this example, by reducing the time delay between the femtosecond laser pulse, generating the filament, and the microwave heater, the plasma density at the instant when the heating starts is increased to 10^{15} cm^{-3}. This increase in the plasma density leads to the reduction in the peak value of population inversion by a factor of about 3. Additionally, the lifetime of population inversion is dramatically reduced by collisional de-excitation and ionization, as shown in Fig. 5.19a. As a result, a much shorter in duration and 50% less intense lasing pulse is generated (Fig. 5.19b).

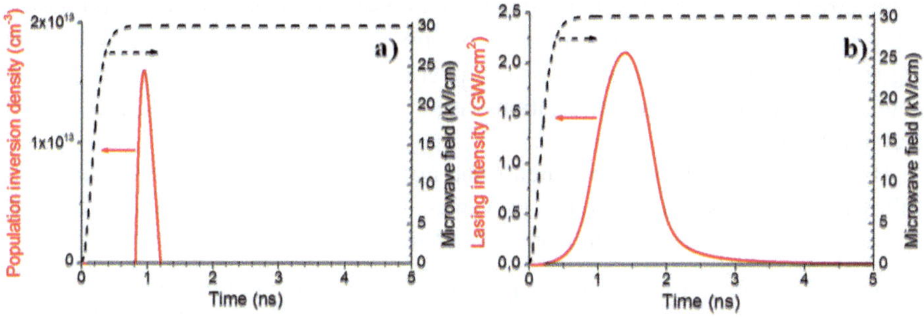

Fig. 5.19 Development of lasing with initial plasma density 10^{15} cm^{-3}. (**a**) Temporal evolution of population inversion density (*solid line*) and (**b**) the lasing pulse at 337 nm wavelength (*solid line*). The *dashed lines* show the temporal profile of the microwave pulse.

The estimated total energy in the lasing pulse is the energy propagating in both forward and backward directions. The partitioning of this energy between forward and backward directions depends on the geometry of the beam crossing between the filament and the microwave radiation. As discussed above, the femtosecond laser pulse, which generates the filament, propagates like a bullet, generating plasma at the position of its current location and leaving a plasma channel behind. If the microwave heater beam is co-propagating with the filament, the beginning of the filament is heated earlier than the end, resulting in more favorable conditions for amplification of the forward-propagating laser pulse. The same is true for the transverse microwave heating geometry. In that case, the same microwave field exists along the filament length. However, the filament is formed not instantaneously but point by point. Thus, the beginning of the filament will be generated and, correspondingly, heated earlier than the end. Sophisticated schemes of beam crossing utilizing a microwave pulse with a tilted wavefront, similar to the schemes used in X-ray lasers [60], will be necessary to enable favorable conditions for amplification in the backward direction.

5.6 Discussion and Conclusions

In this chapter, we have presented an overview of experimental results and models, existing to date, on the standoff laser generation from the second positive system of molecular nitrogen in plasmas of femtosecond filaments. In spite of numerous attempts, no convincing experimental demonstration has been reported so far of the realization of the ultimate goal – an efficient backward-propagating nitrogen lasing in the atmosphere. At the same time, the baseline parameters of the sky laser, which would make it applicable to the standoff nonlinear spectroscopy of the atmosphere, have been discussed in [61]. In that paper, emulations of the atmospheric nitrogen laser with picosecond and nanosecond pulse durations have been generated through parametric frequency conversion and used, in combination with a tunable high-power OPA, for the detection of different gases by stimulated Raman scattering. It has been argued that a μJ-level, nanosecond atmospheric nitrogen lasing can provide reasonable sensitivity, if used in combination with a wavelength-tunable, UV, picosecond, and multi-mJ pump laser.

Based on the results of experiments and numerical simulations, we highlight the following obstacles that, so far, had prevented us from the generation of a backward-propagating coherent emission from atmospheric nitrogen:

- Accurate control over plasma temperature in femtosecond filaments is of a critical importance for achieving lasing.
- The short gain lifetime and the traveling-wave pumping regime make backward-propagating generation directly from a femtosecond filament practically impossible.
- Coherent emissions resulting from supercontinuum and harmonic generation in the filament, which are forward-propagating, cause significant difficulties for achieving backward-propagating lasing, as they provide a relatively intense seed for forward-directed amplification.

As discussed above, polarization and wavelength of the femtosecond pulse are the two parameters that control the electron temperature in the filament plasma. It is an interesting question what kind of a femtosecond laser pulse would enable the most efficient control of the plasma parameters. The use of polarization shaping in the case of long-distance propagation under realistic atmospheric conditions may be problematic, e.g., due to polarization instabilities in filamentation [6]. As far as using a long-wavelength femtosecond laser source for pumping nitrogen air lasing is concerned, low plasma density in mid-IR filaments may pose a problem. From the general $\propto\lambda^2$ scaling of plasma refraction, a significantly (by an order of magnitude) lower plasma density is expected in a mid-IR laser filament compared to that in the near-IR case. Mid-infrared filamentation is an active research field by itself. A potential contribution of higher-order nonlinearities, as well as the effects of temporal self-steepening may additionally reduce plasma generation [62], but those are still open questions.

New possibilities for air lasing may open up in the future through the application of a new and not yet well-studied regime of laser filamentation – superfilamentation [63]. Superfilamentation is the propagation regime of a laser pulse with the peak power several orders of magnitude above the critical power for self-focusing, under the conditions of relatively tight external beam focusing. Another promising approach, which has not been extensively investigated, relies on the use of high peak power picoseconds laser pulses for pumping air lasing. The first experiment of that kind has been recently reported [24]. Picosecond laser filamentation closely mimics the conditions in an impulsive electrical discharge and allows for the direct Joule heating of plasma to the temperature necessary for efficient collisional pumping.

Igniter-heater concept is very promising as it may offer a solution for the problem of the traveling-wave excitation in a plasma filament. Plasma refraction for the optical heater beam may pose a problem, and more research is needed in order to pinpoint the optimum pumping configuration. Microwave heating offers a promising alternative to the heating by laser pulses. Implementing microwave heating at high altitude in the atmosphere may circumvent the requirement of very high peak field of the microwave heater, as the breakdown threshold field is approximately proportional to the ambient pressure.

In conclusion, although the idea of the optically pumped, backward-propagating remote nitrogen lasing in air has been suggested more than a decade ago, this proposition has not been yet materialized. This outstanding unresolved problem in modern laser science has both a tremendous practical potential and many interrelated facets that are interesting from the fundamental physics perspective.

References

1. P.R. Hemmer, R.B. Miles, P. Polynkin, T. Siebert, A.V. Sokolov, P. Sprangle, M.O. Scully, Standoff spectroscopy via remote generation of a backward-propagating laser beam. PNAS **108**, 3130 (2011)
2. L. Yuan, A.A. Lanin, P.K. Jha, A.J. Traverso, D.V. Voronine, K.E. Dorfman, A.B. Fedotov, G.R. Welch, A.V. Sokolov, A.M. Zheltikov, M.O. Scully, Coherent Raman Umklappscattering. Laser Phys. Lett. **8**, 736 (2011)
3. A. Braun, G. Korn, X. Liu, D. Du, J. Squier, G. Mourou, Self-channeling of high-peak-power femtosecond laser pulses in air. Opt. Lett. **20**, 73 (1995)
4. S.L. Chin, *Femtosecond Laser Filamentation*, Springer Series on Atomic, Optical and Plasma Physics 55 (Springer Science+Business Media, LLC, New York, 2010)
5. M. Durand, A. Houard, B. Prade, A. Mysyrowicz, A. Durécu, B. Moreau, D. Fleury, O. Vasseur, H. Borchert, K. Diener, R. Schmitt, F. Théberge, M. Chateauneuf, J.-F. Daigle, J. Dubois, Kilometer range filamentation. Opt. Express **21**, 26836 (2013)
6. A. Couairon, A. Mysyrowicz, Femtosecond filamentation in transparent media. Phys. Rep. **441**, 47 (2007)
7. H.G. Heard, Ultra-violet gas laser at room temperature. Nature **200**, 667 (1963)
8. J. Stebbins, A.E. Whitford, P. Swings, A strong infrared radiation from molecular nitrogen in the night sky. Astrophys. J. **101**, 39 (1945)
9. R.W. Dreyfus, R.T. Hodgson, Electron beam excitation of the nitrogen laser. Appl. Phys. Lett. **20**, 195 (1972)
10. N.G. Basov, A.N. Brunin, V.A. Danilychev, V.A. Dolgikh, O.M. Kerimov, A.N. Lobanov, S.I. Sagitov, A.F. Suchkov, High-pressure ultraviolet laser utilizing Ar+N_2 mixture. Sov. J. Quant. Electron. **5**, 1218 (1976)
11. R.S. Kunabenchi, M.R. Gorbal, M.I. Savadatt, Nitrogen lasers. Progr. Quant. Electr. **9**, 259 (1984)
12. V.A. Vaulin, V.N. Slinko, S.S. Sulakshin, Air ultraviolet laser excited by high-power microwave pulses. Sov. Journ. Quant. Electr. **18**, 1457 (1988)
13. M. Alden, U. Westblom, J.E.M. Goldsmith, Two-photon-excited stimulated emission from atomic oxygen in flames and cold gases. Opt. Lett. **14**, 305 (1989)
14. A. Dogariu, J.B. Michael, M.O. Scully, R.B. Miles, High-gain backward lasing in air. Science **331**, 442 (2011)
15. A. Laurain, M. Scheller, P. Polynkin, Low-threshold bidirectional air lasing. Phys. Rev. Lett. **113**, 253901 (2014)
16. A. Dogariu, R.B. Miles, in *Frontiers in Optics 2013/Laser Science XXIX*, (Laser Science, Orlando, 2013)
17. Q. Luo, W. Liu, S.L. Chin, Lasing action in air induced by ultra-fast laser filamentation. Appl. Phys. B Lasers Opt. **76**, 337 (2003)

18. J.H. Marburger, Self-focusing: theory. Prog. Quantum Electr. **4**, 35 (1975)
19. D. Kartashov, S. Ališauskas, G. Andriukaitis, A. Pugžlys, M. Shneider, A. Zheltikov, S.L. Chin, A. Baltuška, Free-space nitrogen gas laser driven by a femtosecond filament. Phys. Rev. A **86**, 033831 (2012)
20. W.R. Bennett, W.L. Faust, R.A. McFarlane, C.K.N. Patel, Dissociative excitation transfer and optical maser oscillation in Ne-O_2 and Ar-O_2 rf discharges. Phys. Rev. Lett. **8**, 470 (1962); W.R. Bennett Jr., Optical spectra excited in high pressure noble gases by alpha impact. Ann. of Phys. **18**, 367 (1962)
21. G. Andriukaitis, D. Kartashov, D. Lorenc, A. Pugžlys, A. Baltuška, L. Giniūnas, R. Danielius, J. Limpert, T. Clausnitzer, E.-B. Kley, A. Voronin, A. Zheltikov, Hollow-fiber compression of 6 mJ pulses from a continuous-wave diode-pumped single-stage Yb; Na:CaF2 chirped pulse amplifier. Opt. Lett. **36**, 1914 (2011)
22. G. Andriukaitis, T. Balčiūnas, S. Ališauskas, A. Pugžlys, A. Baltuška, T. Popmintchev, M.-C. Chen, M.M. Murnane, H.C. Kapteyn, 90 GW peak power few-cycle mid-infrared pulses from an optical parametric amplifier. Opt. Lett. **36**, 2775 (2011)
23. S.V. Pancheshnyi, S.M. Starikovskaia, A.Yu. Starikovskii, Measurements of rate constants of the $N_2(C^3\Pi_u\ \upsilon'=0)$ and $N_2^+(B^2\Sigma_u^+\ \upsilon'=0)$ deactivation by N_2, O_2, H_2, CO and H_2O molecules in afterglow of the nanosecond discharge. Chem. Phys. Lett. **294**, 523 (1998)
24. D. Kartashov, S. Ališauskas, A. Baltuška, A. Schmitt-Sody, W. Roach, P. Polynkin, Remotely pumped stimulated emission at 337 nm in atmospheric nitrogen. Phys. Rev. A **88**, 041805(R) (2013)
25. https://jlf.llnl.gov.
26. J. Kasparian, J.-P. Wolf, Physics and applications of atmospheric nonlinear optics and filamentation. Opt. Express **16**, 466 (2008)
27. D. Kartashov, Nitrogen laser from a filament Symposium. Femtosecond Filamentation and Standoff Laser Sensing, Vienna, 28–29 March 2012
28. D. Kartashov, S. Ališauskas, A. Pugžlys, M.N. Shneider, A. Baltuška, Theory of a filament initiated nitrogen laser. J. Phys. B **48**, 094016 (2015)
29. D. Kartashov, S. Ališauskas, A. Pugžlys, A. Voronin, A. Zheltikov, M. Petrarca, P. Béjot, J. Kasparian, J.-P. Wolf, A. Baltuška, Mid-infrared laser filamentation in molecular gases. Opt. Lett. **38**, 3194 (2013)
30. S. Mitryukovskiy, Y. Liu, P. Ding, A. Houard, A. Mysyrowicz, Backward stimulated radiation from filaments in nitrogen gas and air pumped by circularly polarized 800 nm femtosecond laser pulses. Opt. Express **22**, 12750 (2014)
31. S. Mitryukovskiy, Y. Liu, P. Ding, A. Houard, A. Couairon, A. Mysyrowicz, Plasma luminescence from femtosecond filaments in air: evidence for impact excitation with circularly polarized light pulses. Phys. Rev. Lett. **114**, 063003 (2015)
32. P. Ding, S. Mitryukovskiy, A. Houard, E. Oliva, A. Couairon, A. Mysyrowicz, Y. Liu, Backward lasing of air plasma pumped by circularly polarized femtosecond pulses for the saKe of remote sensing (BLACK). Opt. Express **22**, 29964 (2014)
33. J. Yao, H. Xie, B. Zeng, W. Chu, G. Li, J. Ni, H. Zhang, C. Jing, C. Zhang, H. Xu, Y. Cheng, Z. Xu, Gain dynamics of a free-space nitrogen laser pumped by circularly polarized femtosecond laser pulses. Opt. Express **16**, 19005 (2014)
34. D. Kartashov, S. Ališauskas, G. Andriukaitis, A. Pugžlys, M. Shneider, B. Landgraf, A. Hoffmann, P. Polynkin, C. Spielmann, A. Baltuska, Filament initiated standoff nitrogen laser: theory and experiment. Book of Abstracts COFIL2014, Shanghai, 2014
35. H.L. Xu, A. Azarm, J. Bernhardt, Y. Kamali, S.L. Chin, The mechanism of nitrogen fluorescence inside a femtosecond laser filament in air. Chem. Phys. **360**, 171 (2009)
36. I.A. Kossyi, A.Y. Kostinsky, A.A. Matveyev, V.P. Silakov, Kinetic scheme of the nonequlibrium discharge in nitrogen oxygen mixtures. Plasm. Sour. Science Tech. **1**, 207 (1992)
37. J. Henriques, E. Tatarova, V. Guerra, C.M. Ferreira, Wave driven N2–Ar discharge. I. Self-consistent theoretical model. J. Appl. Phys. **91**, 5622 (2002)

38. M. Moravej, X. Yang, M. Barankin, J. Penelon, S.E. Babayan, R.F. Hicks, Properties of an atmospheric pressure radio-frequency argon and nitrogen plasma. Plasma Sources Sci. Tech. **15**, 204 (2006)
39. K.S. Klopovsky, A.V. Mukhovatova, A.M. Popov, N.A. Popov, O.B. Popovicheva, T.V. Rakhimova, Kinetics of metastable states in high-pressure nitrogen plasma pumped by high-current electron beam. Journ. Phys. D **27**, 1399 (1994)
40. L.V. Keldysh, Ionization in the field of a strong electromagnetic wave. Zh. Eksp. Teor. Fiz. **47**, 1945 (1964); Sov. Phys. JETP **20**, 1307 (1965)
41. A.M. Perelomov, V.S. Popov, and M.V. Terent'ev, Ionization of atoms in an alternating electric field. Sov. Phys. JETP **23**, 924 (1966); Zh. Eksp. Teor. Fiz. **50**, 1393 (1966)
42. V.D. Mur, S.V. Popruzhenko, V.S. Popov, Energy and momentum spectra of photoelectrons under conditions of ionization by strong laser radiation (the case of elliptic polarization). JETP **92**, 777 (2001)
43. P.B. Corkum, N.H. Burnett, F. Brunel, Above-threshold ionization in the long-wavelength limit. Phys. Rev. Lett. **62**, 1259 (1989)
44. Y. Itikawa, Cross sections for electron collisions with nitrogen molecules. Journ. Phys. Chem. Ref. Data **35**, 31 (2006)
45. S. Bodrov, V. Bukin, M. Tsarev, A. Murzanev, S. Garnov, N. Aleksandrov, A. Stepanov, Plasma filament investigation by transverse optical interferometry and terahertz scattering. Opt. Express **19**, 6829 (2011)
46. N.L. Aleksandrov, S.B. Bodrov, M.V. Tsarev, A.A. Murzanev, Y.A. Sergeev, Y.A. Malkov, A.N. Stepanov, Decay of femtosecond laser-induced plasma filaments in air, nitrogen, and argon for atmospheric and sub-atmospheric pressures. Phys. Rev. E **94**, 013204 (2016)
47. P. Sprangle, J. Peñano, B. Hafizi, D. Gordon, M. Scully, Remotely induced atmospheric lasing. Appl. Phys. Lett. **98**, 211102 (2011)
48. J. Penãno, P. Sprangle, B. Hafizi, D. Gordon, R. Femsler, M. Scully, Remote lasing in air by recombination and electron impact excitation of molecular nitrogen. J. Appl. Phys. **111**, 033105 (2012)
49. Y.P. Raizer, *Gas Discharge Physics* (Springer, New York, 1991)
50. E.T. Gerry, Pulsed-molecular-nitrogen laser theory. Appl. Phys. Lett. **7**, 6 (1965)
51. L. Piper, State to state $N_2(A^3\Sigma_u^+)$ energy pooling reactions. I. The formation of $N_2(C^3\Pi_u)$ and the Herman infrared system. Journ. Chem. Phys. **88**, 231 (1988); State to state $N_2(A^3\Sigma_u^+)$ energy pooling reactions. II. The formation and quenching of $N_2(B^3\Pi_g$, v'=1–12). Journ. Chem. Phys. **88**, 6911 (1988)
52. R.T. Brown, D.C. Smith, Optically pumped electric discharge UV laser. Appl. Phys. Lett. **24**, 236 (1974)
53. M. Scheller, N. Born, W. Cheng, P. Polynkin, Channeling the electrical breakdown of air by optically heated plasma filaments. Optica **1**, 125 (2014)
54. J. Papeer, M. Botton, D. Gordon, P. Sprangle, A. Zigler, Z. Henis, Extended lifetime of high density plasma filament generated by a dual femtosecond–nanosecond laser pulse in air. New J. Phys. **16**, 123046 (2014)
55. D. Kartashov, M.N. Shneider, Femtosecond filament initiated, microwave heated standoff nitrogen laser. Journal of Appl. Phys. **121**, 113303 (2017)
56. M.N. Shneider, A. Baltuska, A.M. Zheltikov, Population inversion of molecular nitrogen in an Ar: N_2 mixture by selective resonance-enhanced multiphoton ionization. J. Appl. Phys. **110**, 083112 (2011)
57. F. Théberge, W. Liu, P.T. Simard, A. Becker, S.L. Chin, Plasma density inside a femtosecond laser filament in air: strong dependence on external focusing. Phys. Rev. E **74**, 036406 (2006)
58. A.W. Ali, A.C. Kolb, A.D. Anderson, Theory of the pulsed molecular nitrogen laser. Appl. Opt. **6**, 2115 (1967)
59. V.F. Papakin, A.Y. Sonin, Measurement of the gain of an ultraviolet nitrogen laser. Sov. J. Quant. Electron. **15**, 581 (1985)

60. V.N. Shlyaptsev, P.V. Nickles, T. Schlegel, M.P. Kalachnikov, A.L. Osterheld, Tabletop x-ray laser pumped with subnanosecond and picosecond pulses. Proc. SPIE **2012**, 111 (1993)
61. P.N. Malevich, D. Kartashov, Z. Pu, S. Ališauskas, A. Pugžlys, A. Baltuška, L. Giniūnas, R. Danielius, A.A. Lanin, A.M. Zheltikov, M. Marangoni, G. Cerullo, Ultrafast-laser-induced backward stimulated Raman scattering for tracing atmospheric gases. Opt. Express **20**, 18784 (2012)
62. P. Panagiotopoulos, P. Whalen, M. Kolesik, J.V. Moloney, Super high power mid-infrared femtosecond light bullet. Nature Phot. **9**, 543 (2015)
63. G. Point, Y. Brelet, A. Houard, V. Jukna, C. Milián, J. Carbonnel, Y. Liu, A. Couairon, A. Mysyrowicz, Superfilamentation in air. Phys. Rev. Lett. **112**, 223902 (2014)

Chapter 6
Filament-Assisted Impulsive Raman Spectroscopy

Johanan H. Odhner and Robert J. Levis

6.1 Introduction and Review of Prior Works

This chapter explores the use of nonlinear spectroscopy as a means to detect molecules in the gas phase. This investigation is motivated by the need for remote sensing and provides concepts that can incorporate backward lasing schemes. The current state of the art with respect to remote Raman sensing using spontaneous Raman is reviewed, and then the use of stimulated Raman and impulsive Raman is presented. We review the use of remote Raman spectroscopy with potential applications in atmospheric sensing, pollution monitoring, and chemical signature detection including explosives, radiological signatures, and biological pathogens.

Coherent optical probing of the vibrational and rotational modes available when two or more atoms are bound together to form a molecule represents a means to classify the molecules in a given volume. There are 3 N-5 and 3 N-6 available vibrational normal modes for linear and nonlinear molecules, respectively, where N is the number of atoms. Thus, in larger molecules of interest for remote sensing, there can be a significant number of vibrational modes available for sensing in one- or two-photon excitation schemes. In the simplest case of a diatomic molecule, there is only one vibrational mode available to probe. In the case of a homonuclear diatomic molecule made up of only one type of atom, the vibrational and rotational modes can only be excited through a two-photon Raman transition. The probability for a Raman transition is proportional to the change in polarizability, α, of the molecule as a function of bond length x, $\mathrm{d}\alpha/\mathrm{d}x$. For heteronuclear diatomics (made up of two different types of atoms), only one-photon vibrational and rotational transitions are allowed. The probability of a transition occurring in this case is given by $\mathrm{d}\mu/\mathrm{d}x$, where μ is the dipole moment of the molecule.

J.H. Odhner • R.J. Levis (✉)
Center for Advanced Photonics Research, Department of Chemistry, Temple University, Philadelphia, PA 19122, USA
e-mail: rjlevis@temple.edu

P. Polynkin, Y. Cheng (eds.), *Air Lasing*, Springer Series in Optical Sciences 208, https://doi.org/10.1007/978-3-319-65220-7_6

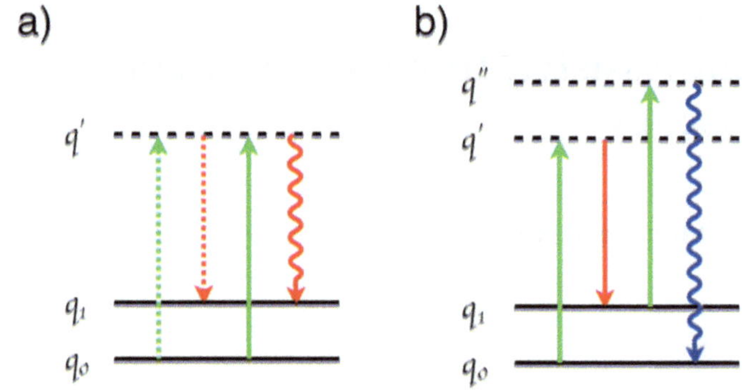

Fig. 6.1 Wave-mixing energy-level diagrams for (**a**) Raman scattering and (**b**) coherent anti-Stokes Raman scattering. The solid (*dotted*) lines represent ket (*bra*) interactions with the final (*wavy*) arrow denoting the coherent emission. The diagrams are read from *left* to *right*

The selection rules govern the experimental possibilities for laser sensing of gas-phase species. In the case of one-photon vibrational excitation, there is the possibility of performing a direct absorption experiment, except that trace species will result in a small absorption on a rather intense probe pulse. For this reason, there are no examples of such experiments being used for remote sensing. In the case of two-photon excitation, a vibrational coherence can be prepared at frequency ω_v that can be probed using a second pulse at frequency ω_o that will result in the formation of sidebands at $\omega_o \pm \omega_v$. These sidebands have only solar background to contend with, and hence very sensitive detection schemes may be developed. Spontaneous Raman spectroscopy could also be employed for remote sensing, but this approach would not take advantage of the directed radiation available with a backward lasing beam. In the case of spontaneous Raman, the signal strength decays as $1/R^2$, where R is the distance from the sample to the detector. In the case of signal imprinted on the probe lasing beam, there is no such dependence.

Raman spectroscopy has been explored as a means of remotely detecting substances since the inception of the laser, due to its inherent ability to sense chemical structure. Raman scattering occurs upon a change in the polarizability of the electron cloud of a molecule due to changes in the relative positions of the constituent atoms. Because each molecule is composed of atoms arranged in a specific manner, each molecule will have a unique Raman spectrum that corresponds to the vibrational and rotational motions of that molecule. Raman spectroscopy is a powerful technique because the molecular specificity allows the determination of chemical signatures. A schematic of the potential Raman excitation schemes is shown in Fig. 6.1. The first diagram in Fig. 6.1a represents Raman scattering between two rovibrational or electronic levels, q_0 and q_1. The upper state, q', represents a virtual state in the case of nonresonant Raman scattering and a real state in the case of resonant Raman scattering. Spontaneous, stimulated, and impulsive Raman scatterings are all represented by the diagram in Fig. 6.1a, depending on the origin of the interacting electric fields and the interaction timing. The second diagram shown in Fig. 6.1b represents coherent anti-Stokes Raman scattering, in which the pump and Stokes pulses, represented by the first two arrows, create a coherence that interacts with the pump electric field (the third arrow) to produce anti-Stokes-shifted Raman emission.

Remote Raman monitoring of atmospheric gases has been under investigation since shortly after the inception of the laser. The first report on the experimental measurement of Raman scattering from nitrogen and oxygen in the atmosphere was in 1967 using a 1 mJ and 10 ns pulsed nitrogen laser coupled to a 20 cm telescope and a photomultiplier tube [1] and was followed closely by a Raman ranging measurement of the atmosphere using Raman scattering from nitrogen to avoid the spectral overlap at the fundamental wavelength from scattering off both gases and aerosols [2]. The potential of remote Raman scattering for atmospheric monitoring and as a molecule-specific and universal detection method was recognized in the following years [3], and two tracks of research emerged: (1) the use of abundant atmospheric gases (nitrogen, oxygen, and water) for remote sensing of the atmosphere itself and (2) the potential of Raman spectroscopy for remote detection of trace pollutants in the atmosphere. The former line of research has seen great success and has been developed to the point where routine monitoring of atmospheric temperature profiles, as well as water, aerosol, and ozone concentrations, is now possible [4]. Initial work on the remote Raman detection of pollutants focused on the detection of molecular signatures such as sulfur dioxide [5–7], methane [8, 9], and carbon dioxide [7], and a mobile Raman instrument for smokestack monitoring was built based on this research [10]. More complex spectroscopic signatures such as smoke from automobile exhaust and oil combustion were also investigated [3], showing the complex signatures that can be captured by Raman spectroscopy. However, the low Raman scattering cross sections for gases made use of spontaneous Raman scattering extremely challenging for monitoring of trace concentration of atmospheric species for routine atmospheric monitoring [11, 12]. Early on the possibility of detecting forward-scattered Raman-shifted light generated through stimulated Raman scattering (SRS) or coherent anti-Stokes Raman scattering (CARS) was proposed [12, 13], but no immediate follow-up work appears to have been done in either of these areas. More recently, an analysis of CARS has highlighted that wave-vector mismatch in the backward direction makes the generation of a remote CARS signal that propagates back toward the laser source impossible [14].

Studies of remote Raman spectroscopy in the 1970s and 1980s focused primarily on the detection of gas-phase molecules for atmospheric sensing and pollution monitoring. In contrast to atmospheric remote Raman monitoring, where the focus is on collection of Raman-scattered signals, investigations of remote Raman detection have focused mainly on signals scattered from solid or liquid samples on surfaces. Angel et al. [15] first applied standoff Raman techniques at an intermediate range for monitoring chemical spills from liquid and semisolid surfaces. In contrast to earlier work, where kilometer-range distances were considered due to the aim of atmospheric monitoring, the design parameters in that investigation focused on detection of solid (various salts) samples at distances in the 10 to 100 of m range.

Their work was extended to a portable Raman instrument using a UV laser (266 and 248 nm) and demonstrated its use for measurement of bulk liquids, films, and liquids deposited on soil [16, 17]. In 2005 a standoff detection with a portable Raman instrument was employed to measure Raman scattering from energetic materials [18], and this technique has continued to be developed as a means of detecting explosives [19, 20]. Standoff Raman spectroscopy has also found applications in the detection of drugs and other chemical compounds [21, 22]. Around the same time, coherent Raman analogues of the spontaneous and resonance Raman techniques, which have been the focus of the majority of studies, were proposed for the detection of bacterial spores [23]. Coherent Raman spectroscopy for standoff detection was realized in 2008 using a single-beam variant of CARS using ultrashort pulses [24, 25].

While technology has significantly improved the ability to use Raman spectroscopy for remote detection of condensed-phase materials, detecting trace molecules in the gas phase (atmosphere) remains a challenge. Air lasing has the potential to rewrite the landscape for remote sensing due to the phenomenon of backward lasing in a laser-produced plasma channel [14]. While significant hurdles remain in producing atmospheric backward lasing that is sufficiently energetic for spectroscopy, important strides have been made in understanding how backward-propagating amplified spontaneous emission (ASE) can be used to remotely probe the atmosphere. In their proof-of-concept implementation of backward stimulated Raman scattering as a technique for remote detection using air lasing, Malevich et al. [26] demonstrated the use of stimulated Raman gain (SRG) in a counter-propagating geometry in the UV region where nitrogen lasing occurs. The proposed scheme for stimulated Raman gain/loss spectroscopy utilizes the backward ASE from excited N_2 and N_2^+ as the Stokes Raman beam, while a high-energy, tunable laser pulse along the forward-propagating path acts as the pump. Using a narrowband picosecond or nanosecond UV pulse to simulate the backward-propagating ASE pulse, Malevich et al. demonstrated SRG in a balanced detection arrangement in air (nitrogen and oxygen) and in methane. This concept was extended in a later publication [27], demonstrating SRG between a Stokes pulse from an OPA and a pump pulse actually generated using backward ASE produced through filamentation of a high-energy mid-infrared pulse in a mixture of argon and nitrogen. These experiments represent important milestones in the use of remotely generated signals for standoff spectroscopy, but further innovation is required in order to realize atmospheric sensing using remote Raman detection. The primary drawback of using narrowband pulses is the requirement of tunability in the exciting laser pulse; in the time required for tuning, the forward-propagating pump pulse in frequency and time fluctuations in the atmosphere could lead to a dispersion of a short-lived species or changes in the sample density. Ideally, the entire vibrational fingerprint of the sample would be measured in a single shot. To this end, we focus here on work that has been carried out over the past 8 years in impulsive Raman spectroscopy using femtosecond laser filamentation as the driving source.

6.2 Femtosecond Laser Filaments for Standoff Detection

The phenomenon of laser filamentation [28, 29] leads to the generation of a white-light continuum due to optical shock formation; it is accompanied by a concomitant self-shortening of the pulse. The filamentation process also encompasses a host of optical phenomena that modify the spatiotemporal distribution of the propagating laser pulse. Laser filamentation is characteristically defined as a dynamic interplay between self-focusing of a powerful ultrafast laser pulse and plasma-induced defocusing caused by ionization of the medium due to the high electric field of the focusing laser pulse [30]. This dynamic interplay enables the generation of long-range plasma channels in the atmosphere and is also responsible for spatial, temporal, and spectral pulse reshaping that results in the generation of a self-shortened laser pulse accompanied by a broad optical continuum spanning from the near ultraviolet to the near infrared. These temporal and spectral properties make femtosecond laser filaments promising for standoff detection of molecules.

The life cycle of an individual filament begins with Kerr-induced self-focusing. The optical Kerr effect is responsible for the intensity dependence of refractive index experienced by a high-power laser pulse. For a Gaussian beam profile, the refractive index gradient caused by the Kerr effect mimics a convex lens, causing the phase front of the pulse to curve. The subsequent increase in the on-axis intensity as energy flows from the outer part of the beam to the center exacerbates the self-focusing effect, leading to ever-increasing on-axis intensity. As the intensity of the pulse increases, catastrophic collapse of the pulse is prevented by ionization of the medium, which produces a plasma. The negative refractive index contribution of the free electrons in the plasma acts in the opposite way as the Kerr effect: the lower refractive index on the optical axis compared to the surrounding area acts like a concave lens, defocusing the pulse. Temporally, the ionization front develops at the leading edge of the pulse, causing the pulse to split as the trailing portion of the pulse experiences the defocusing effect of the plasma generated by the leading edge of the pulse [31]. Provided that there is sufficient energy in the back part of the pulse after initial focusing and plasma generation, the trailing (defocused) portion of the pulse energy can undergo Kerr self-focusing once again in the wake of the leading sub-pulse that remains on axis after plasma generation. Upon refocusing, space-time focusing and self-steepening cause an optical shock to occur at the back of the pulse, leading to rapid blue-side spectral broadening and the formation of a nearly single-optical-cycle feature [32].

The temporal feature created during filamentation can be significantly shorter than the motion of any molecular mode in the propagation medium, leading to impulsive Raman excitation of all vibrational (and rotational) modes of the medium [33] until dispersion lengthens the pulse beyond the characteristic time scale of the molecular motions being excited. The impulsive vibrational excitation of molecules in the medium creates a coherent ensemble that persists in the medium until collisions between molecules randomize the phases of the oscillators; whereupon, the macroscopic coherence is dissipated. The macroscopic polarization that results from the collective coherent motion increases the probability for Raman scattering by many orders of magnitude in comparison with spontaneous Raman scattering, thus opening the possibility of gas-phase sensing.

6.3 Filament-Assisted Impulsive Raman Spectroscopy

To enable detection of all potential signature molecules of interest, a robust, ultrashort (<8 fs) excitation pulse is required to impulsively prepare a coherence that can be imprinted on a second probe beam. As discussed above, fs laser filamentation is of considerable interest as such a pump source. Time-resolved filament-based impulsive Raman spectroscopy was first demonstrated in 2008 by Calegari et al. [34]. The white light produced by filamentation in argon was recompressed using chirped mirrors and directed into a gas-phase sample of H_2. The coherent vibrational motion of H_2 excited impulsively by the pump pulse induced a spectral modulation on a probe beam through the time-dependent change in the refractive index, measured by recording the spectrum as a function of the pump-probe time delay. A Fourier transform of the time-dependent response provided the Raman spectrum. The authors concluded that the pump and probe pulses likely underwent additional filamentation in the H_2 gas because the H_2 vibrational stretch detected has a shorter vibrational period (~8 fs) than the measured duration of the pump and probe pulses (10 fs) after compression and before the sample.

Filament-assisted impulsive Raman spectroscopy directly in the wake of a filament produced with a longer (50 fs) driving pulse was demonstrated the following year [35]. In the latter implementation, a 1.5 ps, 400 nm pulse was used to probe the vibrational motion of the constituents of air (nitrogen and oxygen) and hydrogen (see Fig. 6.2). With this longer duration probe, no scanning of delay is required, and the probe acquires sidebands that are Raman shifted. In this hybrid fs-ps geometry, the vibrational Raman spectrum was measured in a single shot, and direct pulse shortening in the filament channel (without external compression) was demonstrated by the observation of the H_2 vibration at ~4155 cm^{-1}. The demonstration of few-cycle pulse formation directly in air is of great consequence for remote sensing, as the requirements of a noble gas medium and post-filament compression was lifted. The use of a low-energy (<2 μJ), picosecond pulse at 400 nm is very similar to the reported parameters of backward lasing in air [27].

The setup in [35] was used to demonstrate the utility of filament-based impulsive Raman spectroscopy for thermometric applications in extreme environments. In this measurement, the Raman response of N_2 in air was probed as a function of the filament pump/ps probe delay as shown in Fig. 6.3. The oscillations and decay are a manifestation of the preparation of a coherent rovibrational wave packet and the subsequent dephasing of that wave packet. Because no thermal energy is initially deposited into the vibrational degrees of freedom in the system, the decay of the

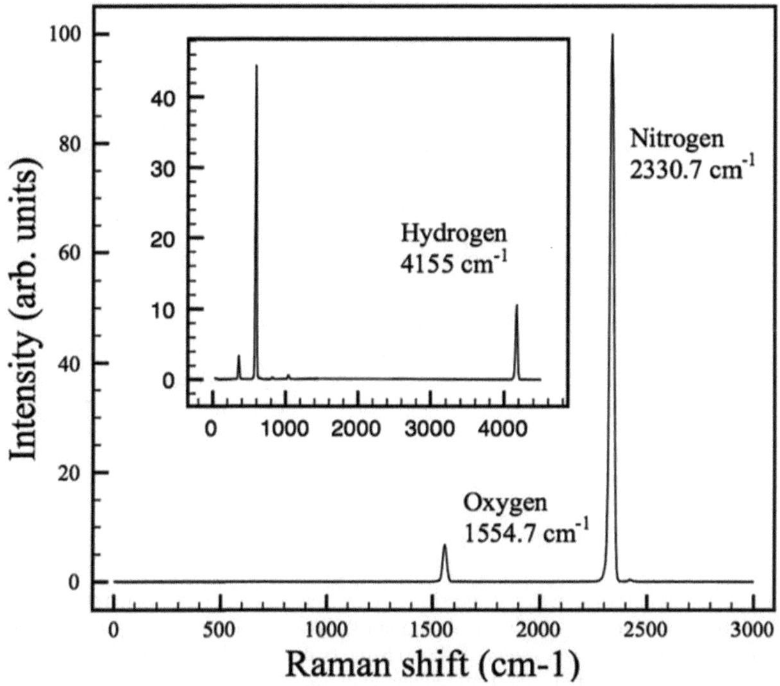

Fig. 6.2 Raman spectrum of air measured using filament-assisted impulsive Raman spectroscopy. Inset: Raman spectrum of hydrogen measured in air showing excitation of both the low-frequency rotational transitions and the high-frequency vibrational mode at ~4155 cm^{-1} (Figure taken from Odhner et al. [35])

rovibrational wave packet reflects the ambient temperature distribution (i.e., ~300 K at room temperature). Measurements in a natural gas/air flame (inset to Fig. 6.3) showed the feasibility of filament-assisted spectroscopy for standoff measurements of flame temperature. These results were validated using a compact analytical expression for the temperature-dependent rovibrational wave packet evolution, which is consistent with results from time-resolved coherent anti-Stokes Raman spectroscopy measurements made using a conventional CARS beam geometry [36].

Because of the relationship between the driving pulse duration and the degree of impulsive Raman excitation of a system, filament-assisted impulsive Raman spectroscopy can be used to map out the dynamics of temporal reshaping that occurs in the filament channel. This is important because the spatial position in the filament at which the Raman excitation of molecular modes is maximized determines the optimal probing position for coherent Raman spectroscopy. In a counter-propagating geometry, the probe beam (generated by backward lasing from a leader filament) will only overlap with the Raman pump pulse for a brief moment in space and time. The overlap is dictated by both the respective pulse durations (or, in the case of a coherently prepared medium, the dispersion time of the rovibrational wave packet) and the timing between the second, forward-propagating laser pulse and the

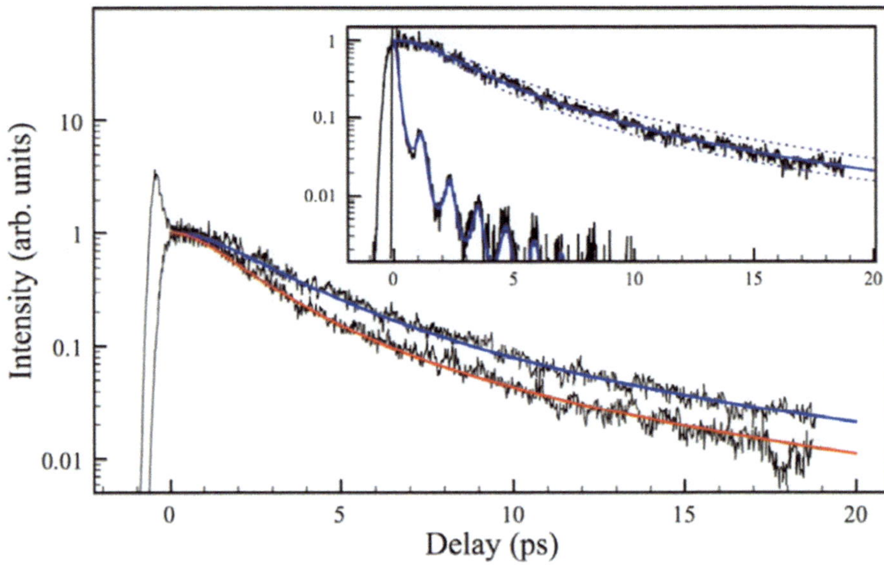

Fig. 6.3 Rovibrational wave packet dispersion of nitrogen (fit, *blue*) and oxygen (fit, *red*) in ambient air (~300 K) overlaid with fits using equation (3) from Odhner et al. [35]. Inset: Nitrogen at room temperature fit with 300 K (*solid*) and ±50 K intervals (*dotted*) and nitrogen in a methane-air flame fitted at 3000 K (Figure taken from Odhner et al. [35])

pulse produced by backward lasing. The Raman excitation profile of a femtosecond laser pulse-induced filament was measured [37] using a modified arrangement of the setup described in reference [35]. The basic concept is to scan the length of a filament produced by a femtosecond pulse using a narrowband probe while recording the Raman spectra at each point along the filament. Since the major ambient species in air, namely, oxygen, nitrogen, and water, have Raman-active vibrational frequencies of 1555, 2331, and 3650 cm^{-1}, corresponding to 21.5, 14.3, and 9.1 fs periods of vibration, it is possible to observe the onset of pulse shortening in the filament channel by the appearance and dynamics of each Raman line. Figure 6.4 shows the dynamics of the Raman lines as a function of distance from the filament-inducing lens along with the fluorescence profile of the filament, which reflects the intensity and is roughly a proxy for the plasma density in the filament channel. Interestingly, for the filamentation conditions used in the experiment, the most significant pulse shortening occurs after the main focusing event, which is consistent with models of filament-induced pulse shortening. The Raman signals for oxygen and nitrogen appear before the linear focus, indicating that initial Kerr self-focusing shortens the on-axis pulse in time through temporal concentration of the intensity onto the optical axis [38]. The appearance of the Raman lines sets an upper limit on the duration of pulse [39], suggesting that even before the geometric focus pulse features as short as 14.3 fs (the vibrational period of nitrogen) are formed. However, it is not until after the first focusing event (region I in Fig. 6.4a) that significant pulse shortening takes place, leading to two orders of magnitude increase in the Raman

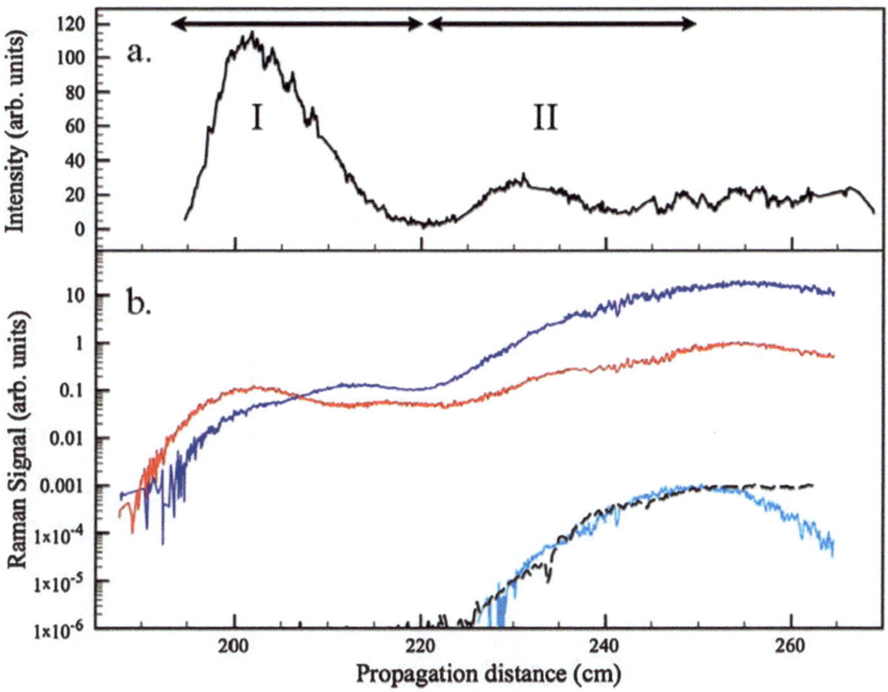

Fig. 6.4 Fluorescence (**a**) and Raman (**b**) profiles measured in a femtosecond filament generated using an 2 mJ, 50 fs pulse as a function of distance from the $f = 2.07$ m lens used to generate the filament. The Raman lines tracked are nitrogen (*dark blue*), oxygen (*red*), and water (*light blue*). The *dashed line* in (**b**) is the integrated blue wing of the filament spectrum measured as described in [37]

signals of nitrogen and oxygen. The appearance of the vibrational mode of water indicates the formation of a sub-10 fs pulse feature. Pulse shortening in region II of Fig. 6.4a is driven by space-time focusing and self-steepening of the rear portion of the pulse, which gently refocuses in the wake of the leading edge and forms an optical shock [40]. This observation is promising for remote Raman sensing, as the ability to probe a coherently prepared medium has significant advantages of increasing signal strength and simultaneous excitation of many mode frequencies. It also highlights a challenge of using filamentation as both a source of plasma for the generation of a backward-propagating laser beam and for vibrational excitation: the spatial overlap between the two pulses must be optimized so that the backward-propagating beam propagates through the coherently excited medium so that Raman sideband formation might be effectively achieved.

Beyond using filament-assisted impulsive Raman spectroscopy (FAIRS) as a means to diagnose the filamentation process itself, we have also extended the FAIRS method as a tool for detecting organic compounds and the reaction products of atmospheric chemical processes. One of the advantages of using filamentation as a source of Raman excitation is the ability to impulsively excite the entire range of

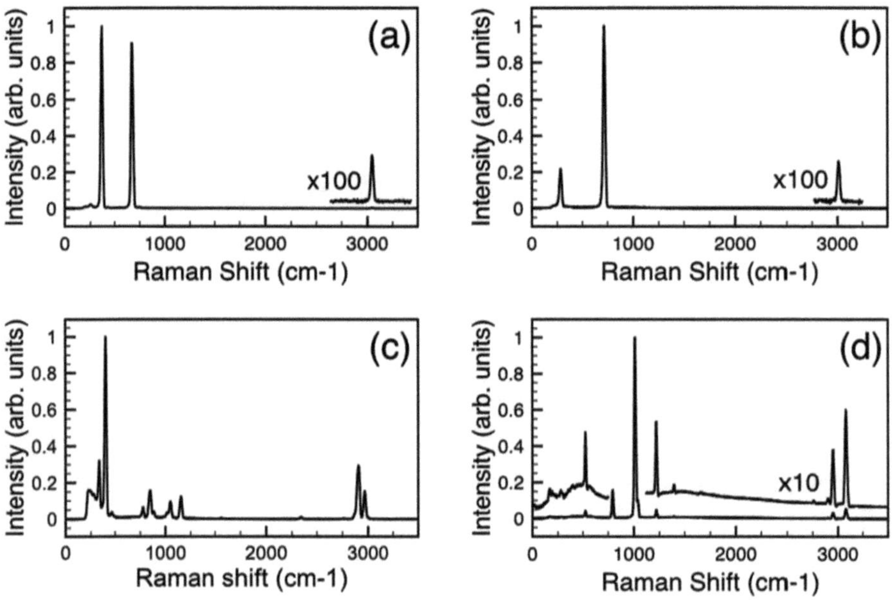

Fig. 6.5 Raman spectra of vapor samples obtained with filament-assisted impulsive Raman spectroscopy: (**a**) chloroform, (**b**) methylene chloride, (**c**) pentane, and (**d**) toluene. (Data from Odhner et al. [41])

vibrational motions, from low-energy bending and twisting modes to the high-frequency hydrogen-stretching modes found in C–H, O–H, and N–H bonds. Other coherent Raman methods typically require scanning the spectral and/or temporal overlap between pump and Stokes laser beams in order to map out the vibrational fingerprint of a region of space. With broadband impulsive excitation, all Raman-active modes can be excited simultaneously and read out by imprinting sidebands on a narrowband pulse. The Raman spectra shown in Figs. 6.5 and 6.6 illustrate this point, demonstrating the detection of several chlorocarbons and hydrocarbons (Fig. 6.5), as well as potential signatures of improvised explosive devices such as nitromethane, ammonia, and gasoline (Fig. 6.6) in the gas phase [41]. The relatively modest laser parameters (50 fs, 2 mJ filament pulse and 0.7 ps, 12 μJ probe pulse) are promising for extension to lower detection limits and more challenging detection geometries.

Excitation of gas-phase signatures can be accomplished at longer distances from the laser as well. Figure 6.7 shows the Raman spectrum of the C–H stretch region of gasoline vapor measured at 7 m and 13 m from the filament-inducing lens with $f = 5$ m and 10 m, respectively. Otherwise, the excitation and detection pulse parameters are the same as that used in Figs. 6.5 and 6.6. The ps probe beam crossed the filament above a cuvette containing the gasoline, and the sidebands were detected after the interaction region.

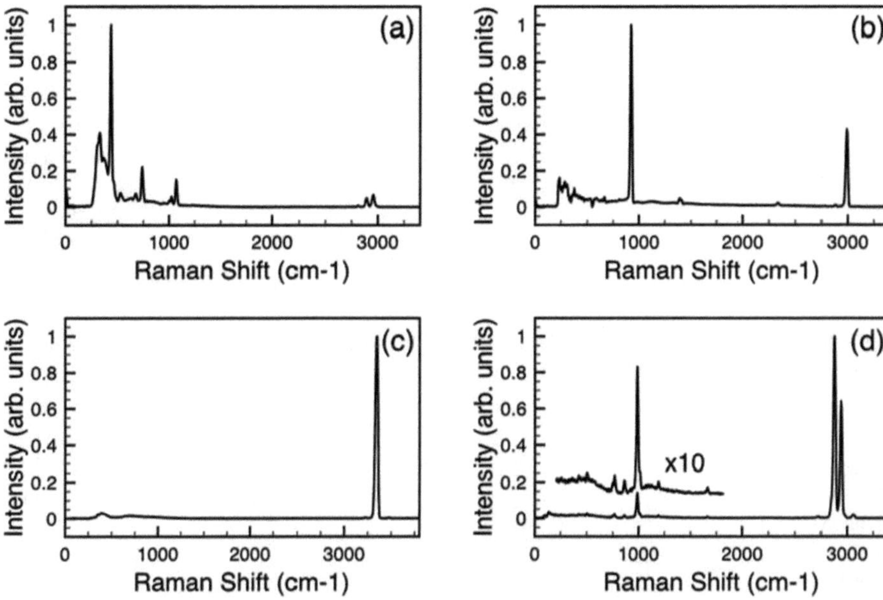

Fig. 6.6 Raman spectra of vapor samples obtained with filament-assisted impulsive Raman spectroscopy. (**a**) Triethylamine, (**b**) nitromethane, (**c**) ammonia, and (**d**) gasoline (Data from Odhner et al. [41])

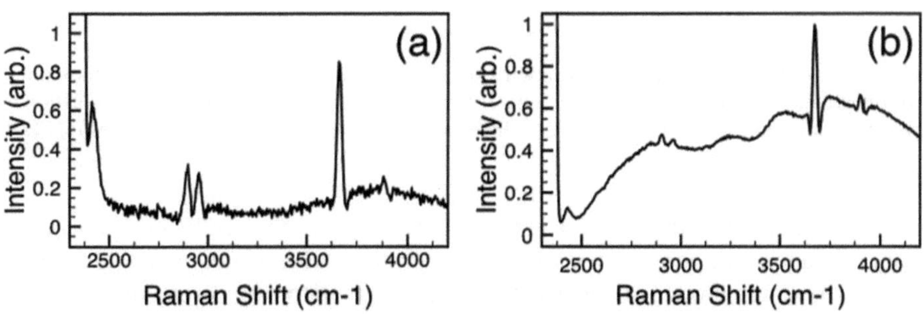

Fig. 6.7 Raman spectra of the gasoline C–H stretch region in humid air at (**a**) 7 m and (**b**) 13 m from the filament-generating lens. The feature at ~3880 cm^{-1} is the cascaded stimulated Raman signal of oxygen (nitrogen) off of the nitrogen (oxygen) Raman line in air

In addition to the detection of stable molecular species, an important area of interest is detecting unstable species or ions. Ozone and nitrogen oxides are important signatures for the presence of radioactive materials and result from the interaction of energetic decay particles with air. Filament-induced impulsive Raman spectroscopy has been applied to the detection of these molecules recently.[3] A sensitive gas-phase Raman spectrometer was constructed using a 4-*f*, zero-dispersion grating compressor as a filter to set the probe bandwidth and energy and a second 4-*f* filter to select an observation window for detection, as shown in Fig. 6.8.

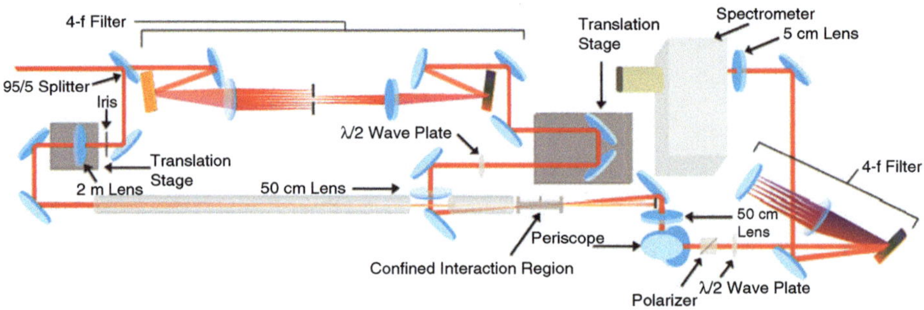

Fig. 6.8 Experimental setup for filament-assisted impulsive Raman spectroscopy of ozone and nitrogen oxides (Taken from Ref. [42])

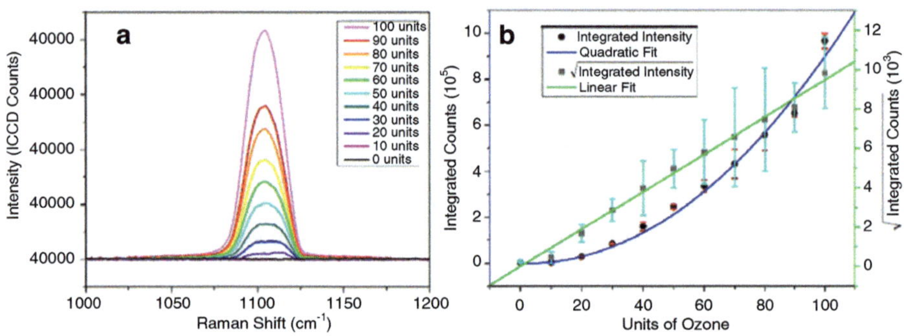

Fig. 6.9 (**a**) Raman signal from the symmetric ozone stretch as a function of ozone generator units. (**b**) Integrated ozone peak intensity and square root of the integrated peak intensity as a function of concentration showing $\propto N^2$ dependence of the coherently generated signal (Figure based on [42])

All frequencies but the desired probe bandwidth are physically blocked in the Fourier plane of the zero-dispersion compressor to provide a high-contrast bandpass filter. Ozone was produced either with an ozone generator (by corona discharge) or by ionization/dissociation and recombination of oxygen atoms within the filament itself. In the ozone generator measurement, the generator output was directed into a sample tee with apertures on the laser inlet and exit ports to maximize the amount of ozone retained in the filament-probe interaction region.

The concentration of ozone was calibrated via absorption measurements of the Hartley band using a UV-Vis spectrometer configured to operate directly in the sample tee. A plot of the integrated impulsive Raman signal intensity as a function of ozone concentration shows the quadratic dependence on the number density, N, that is expected for coherent Raman scattering (shown in Fig. 6.9). Note that an incoherent (spontaneous Raman scattering) measurement would result in a linear signal response with N. A calculation of the limit of detection for the system from these measurements yields a value of ~300 ppm using the equation

$$\mathrm{LOD} = \sqrt{3\sigma}\big/ m, \tag{6.1}$$

where σ is the standard deviation of the detector noise and m is the slope of the linear regression of the response.

Upon increasing the probe energy by opening the slit in the $4-f$ filter that controls the bandwidth of the probe, interference was observed in the Raman signal as shown in Fig. 6.10. This interference provides a second means of identifying Raman signal of signature molecules of interest and has been shown to arise from heterodyning of the Raman-shifted probe spectrum and cross-phase modulation between the pump and the probe beams. The interference as a function of pump-probe delay is shown in Fig. 6.10a where the fringe spacing decreases as a function of increasing time delay. To understand the origin of the fringes, the Raman signal was modeled as an impulsively excited molecular vibration interacting with an arbitrarily shaped probe pulse following [43]:

$$S_{\mathrm{CRS}}(\omega,\tau) = \left| \int_{\infty}^{-\infty} \mathrm{d}t\, e^{i\omega t}\, E_{\mathrm{pr}}(t) \cdot \int_{\infty}^{0} \mathrm{d}t_2 \left| E_{\mathrm{pu}}(t+\tau+t_2) \right|^2 \cdot \left(\chi_{\mathrm{K}}(t_2) + \chi_{\mathrm{vib}}(t_2) \right) \right|^2, \tag{6.2}$$

where $E_{\mathrm{pr}}(t)$ and $E_{\mathrm{pu}}(t)$ are the probe and filament electric fields, respectively, χ_{K} and χ_{vib} are the linear response functions associated with the optical Kerr and vibrational Raman nonlinearities, and τ is the pump-probe time delay. Following [44], the

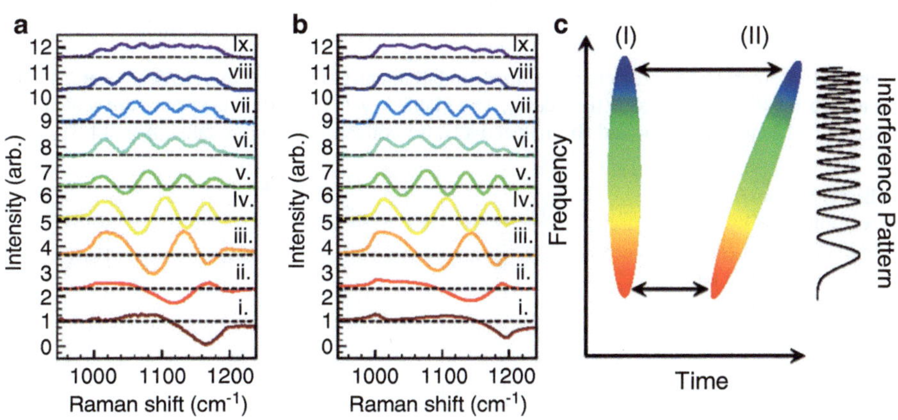

Fig. 6.10 Experimental (**a**) and simulated (**b**) interference in the ozone Raman signal due to heterodyning between filament-induced cross-phase modulation and the Raman-scattered probe light. The time delays shown are (i) 0 fs, (ii) 130 fs, (iii) 266 fs, (iv) 410 fs, (v) 540 fs, (vi) 667 fs, (vii) 800 fs, (viii) 933 fs, and (ix) 1060 fs. (**c**) A cartoon explaining the dispersion in the fringe period due to the different temporal delay between the dispersion-free cross-phase modulation (I) and frequency-chirped Raman-shifted probe pulse (II) (Figure taken from McCole Dlugosz et al. [42])

optical Kerr effect is modeled as a phase modulation that is proportional to the pump temporal profile. The Raman spectrum was calculated using an experimentally retrieved filament pulse profile [45] as well as a simulated bandwidth-limited 10 fs pulse. The Raman line shapes were essentially the same in these two cases. The calculated response is also shown as a function of pump-probe delay (Fig. 6.10b). The calculated response is in excellent agreement with the measured Raman spectra. The heterodyning of the signal through interference with the filament-induced cross-phase modulation can potentially increase the instrument sensitivity considerably.

The detection of ozone formed after the filament pump pulse interacts with air for several minutes is shown in Fig. 6.11. Spectra are included for pump-probe delays of 400 and 540 fs. Superimposed on the spectra are measurements made using the ozone generator with the same pump-probe delays. In this case, we calculate the concentration of the accumulated filament-generated ozone to be seven parts per thousand. Finally, the impulsive Raman spectrum taken for NO_2 is shown in Fig. 6.12. This spectrum was acquired by leaking NO into the sample and allowing reaction with ambient air to occur to produce NO_2.

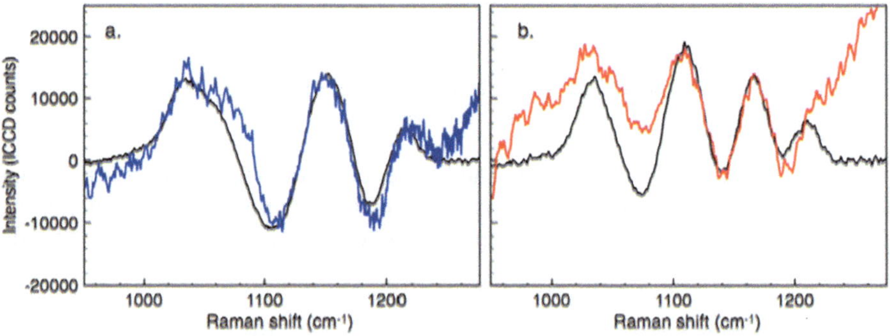

Fig. 6.11 Raman signal of discharge- (*black*) and filament-generated (*blue*, *red*) ozone at (**a**) ~400 fs and (**b**) ~540 fs pump-probe time delay. The filament-generated Raman spectra are scaled by a factor of ×10 and ×18 in (**a**) and (**b**), respectively, for comparison (Figure taken from [42])

Fig. 6.12 Raman spectrum of NO (1875 cm^{-1}) and NO_2 (1316 cm^{-1}) (Data taken from Ref. [42])

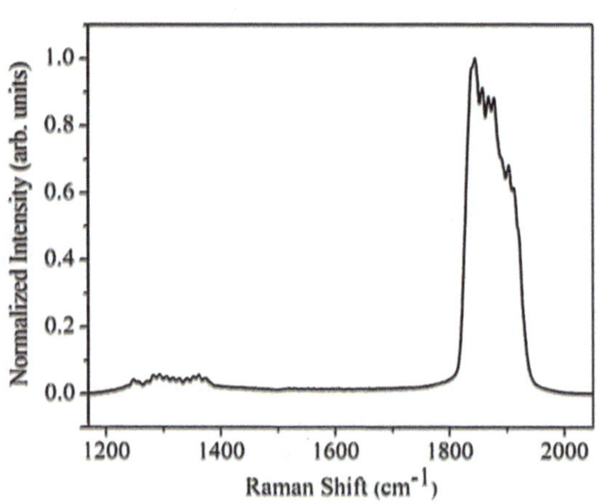

6.4 Rotational Spectroscopy

Rotational Raman spectroscopy can also be used to determine the identity of molecules due to the unique moments of inertia about the principal axes of each molecule. This sets the rotational periods of a given molecule that can be measured in either the spectral or temporal domain. Impulsive rotational Raman excitation exerts a torque on any molecule with an anisotropic polarizability that tends to align the highest polarizability axis with the polarization axis of the laser field. This in turn leads to field-free rotational alignment after the laser-molecule interaction and, subsequently, revivals of the rotational wave packet [46]. The phenomenon has less stringent requirements on pulse duration in comparison with impulsive vibrational excitation because the energy-level spacing is typically much smaller for rotational transitions. Wave-packet revivals can be detected in a variety of ways, but inline sensing modalities such as detecting the induced birefringence [47] or spatial focusing/defocusing [48, 49] induced by the transient refractive index modification on a probe beam by the wave packet might be considered the most promising methods for the application of impulsive rotational Raman excitation to remote detection.

To perform rotational Raman spectroscopy of molecules in air, we have developed a method called spectral-to-temporal amplitude mapping spectroscopy (STAMPS) for rotational revivals [50]. The method is of interest for remote sensing because the rotational revival transient is measured in a single laser shot in comparison with a pump-probe measurement, which can take more than 10^5 laser shots to acquire a spectrum. In the STAMPS measurement, a femtosecond pulse is used to initiate the rotational wave packet with polarization in the vertical direction. The probe pulse consists of a broad bandwidth pulse created through filamentation in air or sapphire, followed by stretching to a duration of more than 50 ps. The polarization of the probe beam is –45° with respect to the pump. After the probe beam interacts with the reviving medium, it is passed through a polarization filter at +45° with respect to the pump polarization. Frequencies in the chirped probe pulse that have experienced some transient birefringence pass through the filter. The transient birefringence occurs upon a revival of the wave packet, thus imprinting the temporal dynamics onto the spectrum of the chirped probe beam in a single laser shot. To perform heterodyned detection, a quarter-wave plate is inserted into the probe beam path before the final polarization analyzer, and two measurements are performed with a small positive or negative rotation. Subtraction of the two measurements provides the heterodyned signal [47, 51].

Figure 6.13 shows the STAMPS measurements of nitrogen, oxygen, and CO_2 [50]. A Fourier transform of the transients provides the rotational spectrum of the molecules with rotational constants in agreement with literature. In these measurements, the rotational revivals of the wave packet are then mapped onto a chirped continuum (400–800 nm) pulse created by laser filamentation in argon. Time-resolved rotational spectra are recorded over a 65 ps time window. The method has

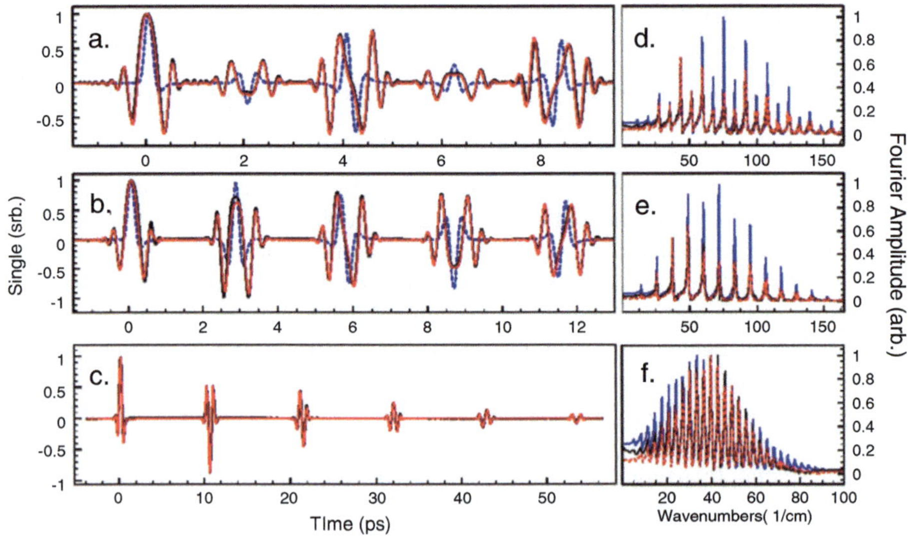

Fig. 6.13 STAMPS measurement of (**a**) nitrogen, (**b**) oxygen, and (**c**) carbon dioxide. The blue dotted lines show the calculated induced-phase modulation due to rotational revivals, while the black (*red*) data shows the experimental (calculated) STAMPS signal. (**d**), (**e**), and (**f**) show the Fourier transform of the signals shown in (**a**), (**b**), and (**c**), respectively (Figure taken from McCole et al. [50])

been extended to asymmetric top molecules including ethylene and methanol as shown in Fig. 6.14. In the case of an asymmetric top, the wave packet rapidly dephases due to the differences in the revival time for the three axes, and, in such cases, a full revival is not observed.

6.5 Conclusions

We have discussed spectroscopic techniques that can advantageously utilize remote air lasing. The measurements presented here demonstrate that gas-phase Raman spectroscopy is possible in the wake of a femtosecond filament formed directly in air. Self-shortening of the filament pulse down to a sub-8-fs duration enables the impulsive Raman excitation of all known Raman-active vibrational modes. The use of a ps probe beam enables the detection of all Raman-active vibrations in a single laser shot. Demonstration of sensitivity at a level of 300 ppm in the gas phase suggests the potential for remote detection of interesting signature molecules for explosives, radioactive materials, and perhaps even biohazards. The latter are quite complex, with up to tens of thousands of atoms in a given molecule, and finding unique vibrational or rotational frequencies for classification will be challenging. Single-shot, transient birefringence spectroscopy of rotational revivals was also presented as a potential method for remote detection.

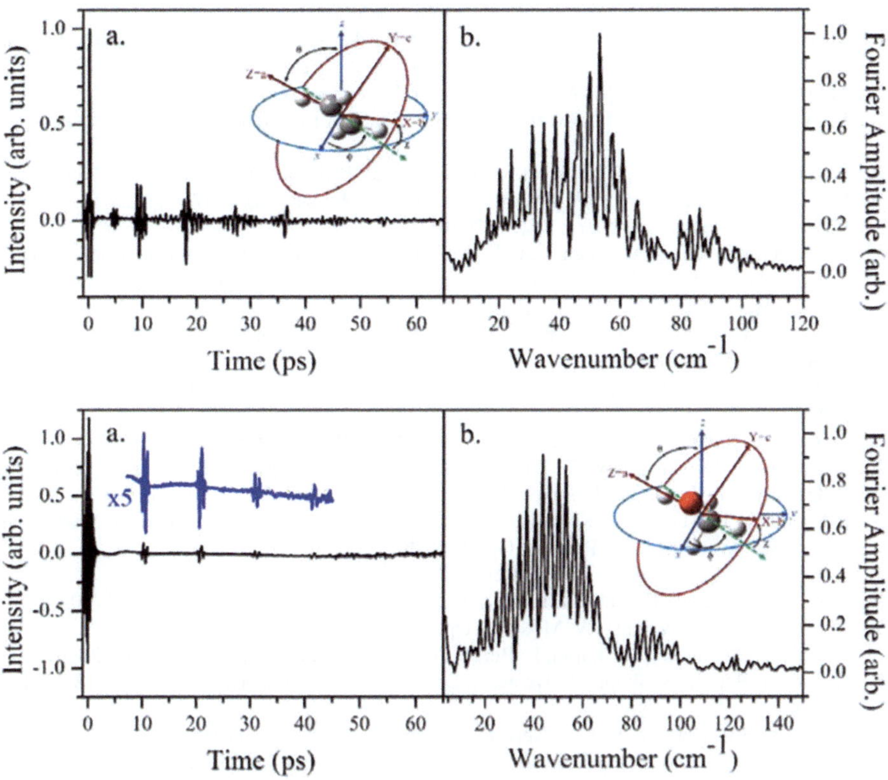

Fig. 6.14 STAMPS measurements of ethylene (*top*) and methanol (*bottom*). (**a**) and (**b**) show the signal and the Fourier transform of the signal, respectively. Insets show the ethylene and methanol molecules in the laboratory frame with respect to their Euler angles (Figure based on [50])

References

1. D.A. Leonard, Nature **216**, 142 (1967)
2. J.A. Cooney, Appl. Phys. Lett. **12**, 40 (1968)
3. H. Inaba, T. Kobayasi, Opto-electronics **4**, 101 (1972)
4. U. Wandinger, *Lidar* (Springer, 2005) Berlin, Germany, p. 241
5. S. Nakahara, K. Ito, S. Ito, A. Fuke, S. Komatsu, H. Inaba, T. Kobayasi, Opto-electronics **4**, 169 (1972)
6. S.K. Poultney, M. Brumfield, J. Siviter, Appl. Opt. **16**, 3180 (1977)
7. T. Kobayasi, H. Inaba, Appl. Phys. Lett. **17**, 139 (1970)
8. J.D. Houston, S. Sizgoric, A. Ulitsky, J. Banic, Appl. Opt. **25**, 2115 (1986)
9. W.S. Heaps, J. Burris, Appl. Opt. **35**, 7128 (1996)
10. S. Nakahara, K. Ito, S. Tamura, M. Kamokiyo, H. Inaba, T. Kobayashi, IEEE J. Quant. Electr. **7**, 325 (1971)
11. H. Kildal, R.L. Byer, Proc. IEEE **59**, 1644 (1971)
12. R.L. Byer, Opt. Quant. Electr. **7**, 147 (1975)
13. J. Fontana, G. Hassin, B. Kincaid, Nature **234**, 292 (1971)
14. P.R. Hemmer, R.B. Miles, P. Polynkin, T. Siebert, A.V. Sokolov, P. Sprangle, M.O. Scully, PNAS **108**, 3130 (2011)

15. S. Angel, T.J. Kulp, T.M. Vess, Appl. Spectr. **46**, 1085 (1992)
16. M. Wu, M. Ray, K.H. Fung, M.W. Ruckman, D. Harder, A.J. Sedlacek, Appl. Spectr. **54**, 800 (2000)
17. M.D. Ray, A.J. Sedlacek, M. Wu, Rev. Sci. Instr. **71**, 3485 (2000)
18. J.C. Carter, S.M. Angel, M. Lawrence-Snyder, J. Scaffidi, R.E. Whipple, J.G. Reynolds, Appl. Spectr. **59**, 769 (2005)
19. S. Wallin, A. Pettersson, H. Östmark, A. Hobro, Anal. Bioanal. Chem. **395**, 259 (2009)
20. K.L. Gares, K.T. Hufziger, S.V. Bykov, S.A. Asher, J. Raman Spectr. **47**, 124 (2016)
21. E.L. Izake, Forensic Sci. Intern. **202**, 1 (2010)
22. A. W. Fountain III, J. A. Guicheteau, W. F. Pearman, T. H. Chyba, S. D. Christesen, in *SPIE Defense, Security, and Sensing* (International Society for Optics and Photonics, 2010), p. 76790H
23. M. Scully, G. Kattawar, R. Lucht, T. Opatrný, H. Pilloff, A. Rebane, A. Sokolov, S. Zubairy, PNAS **99**, 10994 (2002)
24. H. Li, D.A. Harris, B. Xu, P.J. Wrzesinski, V.V. Lozovoy, M. Dantus, M. Opt. Exp **16**, 5499 (2008)
25. O. Katz, A. Natan, Y. Silberberg, S. Rosenwaks, Appl. Phys. Lett. **92**, 171116 (2008)
26. P. Malevich, D. Kartashov, Z. Pu, S. Ališauskas, A. Pugžlys, A. Baltuška, L. Giniūnas, R. Danielius, A. Lanin, A. Zheltikov, Opt. Exp. **20**, 18784 (2012)
27. P. Malevich, R. Maurer, D. Kartashov, S. Ališauskas, A. Lanin, A. Zheltikov, M. Marangoni, G. Cerullo, A. Baltuška, A. Pugžlys, Opt. Lett. **40**, 2469 (2015)
28. A. Couairon, A. Mysyrowicz, Phys. Rep. **441**, 47 (2007)
29. L. Bergé, S. Skupin, R. Nuter, J. Kasparian, J.-P. Wolf, Rep. Prog. Phys. **70**, 1633 (2007)
30. M. Mlejnek, E.M. Wright, J.V. Moloney, Opt. Lett. **23**, 382 (1998)
31. C. Brée, A. Demircan, S. Skupin, L. Berge, G. Steinmeyer, Laser Phys. **20**, 1107 (2010)
32. C. Brée, J. Bethge, S. Skupin, L. Bergé, A. Demircan, G. Steinmeyer, New J. Phys. **12**, 093046 (2010)
33. Y.X. Yan, E.B. Gamble Jr., K.A. Nelson, J. Chem. Phys. **83**, 5391 (1985)
34. F. Calegari, C. Vozzi, S. De Silvestri, S. Stagira, Opt. Lett. **33**, 2922 (2008)
35. J. Odhner, D.A. Romanov, R.J. Levis, Phys. Rev. Lett. **103**, 075005 (2009)
36. T. Lang, M. Motzkus, H. Frey, P. Beaud, J. Chem. Phys. **115**, 5418 (2001)
37. J. Odhner, D. Romanov, R. Levis, Phys. Rev. Lett. **105**, 125001 (2010)
38. O. Kosareva, I. Murtazin, N. Panov, A. Savel'ev, V. Kandidov, S.-L. Chin, Laser Phys. Lett. **4**, 126 (2006)
39. R.A. Bartels, S. Backus, M.M. Murnane, H. Kapteyn, Chem. Phys. Lett. **374**, 326 (2003)
40. C. Brée, A. Demircan, S. Skupin, L. Berge, G. Steinmeyer, Opt. Exp. **17**, 16429 (2009)
41. J.H. Odhner, E.T. McCole, R.J. Levis, J. Phys. Chem. A **115**, 13407 (2011)
42. E.T. McCole Dlugosz, R. Fisher, A. Filin, D.A. Romanov, J.H. Odhner, R.J. Levis, J. Phys. Chem. A **119**, 9272 (2015)
43. S. Mukamel, *Principles of Nonlinear Optical Spectroscopy* (Oxford University Press on Demand, 1999)
44. R. W. Boyd, *Nonlinear Optics* (Academic Press, 2003) Burlington, MA
45. J. Odhner, R.J. Levis, Opt. Lett. **37**, 1775 (2012)
46. T. Seideman, Phys. Rev. Lett. **83**, 4971 (1999)
47. V. Renard, M. Renard, S. Guérin, Y. Pashayan, B. Lavorel, O. Faucher, H.-R. Jauslin, Phys. Rev. Lett. **90**, 153601 (2003)
48. V. Renard, O. Faucher, B. Lavorel, Opt. Lett. **30**, 70 (2005)
49. J. Wu, H. Cai, Y. Tong, H. Zeng, Opt. Exp. **17**, 16300 (2009)
50. E.T. McCole, J.H. Odhner, D.A. Romanov, R.J. Levis, J. Phys. Chem. A **117**, 6354 (2013)
51. V. Loriot, P. Tzallas, E. Benis, E. Hertz, B. Lavorel, D. Charalambidis, O. Faucher, J. Phys. B **40**, 2503 (2007)

Index

A
Air laser, 46
Air lasing
  application, 116
  around-the-corner lasing, 40, 41
  generation mechanisms, 24, 25, 27, 28, 30, 32
  long-wavelength femtosecond laser, 116
  multiphoton pumping, 32–34, 36
  neutral nitrogen molecules, 60
  physical mechanisms, 46
  properties, 22, 24
  remote identification, 85
  signals, 46
Aldén, M., 1–16
Ambient air, 51
Amplification, 51, 52, 98
Amplified spontaneous emission (ASE), 48, 124
Argon
  atmosphere, 37
  concentrations, 20, 39
  preexisting atomic species, 24
  pulse-to-pulse fluctuations, 19, 42
  three-photon excitation, 38
Around-the-corner lasing, 40, 41
Atomic air lasing, 21, 22, 24
Atomic nitrogen, 8
  backward-propagating, 22
  bandpass filters, 40
  density, 34, 35
  dependence, 34
  emission lines, 24
  higher-energy pre-pulse, 34
  laser, 24
Atomic oxygen, 4, 10
  backwards emission, 28
  detector response, 24
  gain region, 26
  hydrogen laser beams, 22
  methane/air flame, 32
  two-photon, 41

B
Backward emission, 49–50
Backward emission signal, 29
Backward-generated SE, 3
Backward laser emission, 27
Backward lasing, 19
Backward-stimulated emission, 46
Baltuska, A., 89–117
Bood, J., 1–16
Brown, M.S., 15

C
Calegari, F., 126
Carbon monoxide, 10–14
Cavityless laser, 20
Cheng, Y., 75–87
Chirped-pulse amplified (CPA), 46
Chu, W., 75–87
Circularly polarized femtosecond pumping, 48–52
Coherent anti-Stokes Raman scattering (CARS), 2, 123
Coherent coupling, 80–82
Combustion diagnostics, 2, 9
Crossed-beam setup, 14, 15

P. Polynkin, Y. Cheng (eds.), *Air Lasing*, Springer Series in Optical Sciences 208, https://doi.org/10.1007/978-3-319-65220-7

D

Ding, P., 45–73
Diode-array detector (DA), 11
Dogariu, A., 16, 19–42

E

Electron collisions
  backward-propagating stimulated emissions, 46–47
  Earth-based receiver, 46
  energy diagram, 47
  filamentation, 46
  forward/backward-propagating, 46
  high-pressure argon gas, 47
  neutral nitrogen molecules, 45
  physical mechanism, 48
  polarized femtosecond laser, 47
  pump pulses, 48
  remote sensing, 46
  superradiance, 46
Electron energy distribution function (EEDF), 103–106

F

Femtosecond filaments
  argon-assisted Bennett mechanism, 97
  backward-directed UV emission, 92
  backward-emitted intensity, 98
  backward-propagating
    lasing, 100
    nitrogen, 115
    seed UV pulse, 98
    stimulated emission, 92
  backward signal detection, 101
  Bennett process, 92
  Bennett pumping scheme, 94
  bichromatic laser generation, 93
  capacitor probe technique, 97
  circular polarization, 98, 101
  COMET laser system, 94
  discharge-pumped laser, 96
  discharge-type nitrogen laser, 94
  experimental investigation, 99
  fluorescence, 92
  forward-propagating lasing, 99
  harmonic radiation, 96
  interpretation, 92
  lasing efficiency, 96
  mid-IR filaments, 93
  molecular nitrogen, 101
  numerical simulations, 94, 116
  OPCPA mid-IR femtosecond system, 97
  picosecond filament, 101
  plasma channel, 95
  population inversion, 99
  quarter-wave plate, 99
  Rayleigh zone, 95
  resonant excitation transfer, 92
  single-shot
    energy, 101
    photograph, 96
    spectrum, 100
  spatial coherence, 94
  spectral diagnostics, 95
  spectral filter, 97
  transmitted UV emission, 95
  two-step kinetic process, 92
  UV emission, 94
Femtosecond laser filaments, 125, 126
Filament-assisted impulsive Raman spectroscopy (FAIRS), 126–130, 133, 134
Following the seed pulse has a width (FWHM), 57
Franck-Condon factors, 11
Frank, J.H., 14
Full width at half maximum (FWHM), 4

G

Gain mechanisms, 48–61
  electron collisions, 45
  neutral nitrogen (*see* Neutral nitrogen)
  picosecond-scale lifetime, 67
Glass filter (GF), 4
Greenhouse gases, 19

H

High-gain lasing
  around-the-corner, 19, 20
  backward-propagating lasing beam, 20
  inert atomic gas, 19
  laser configurations, 20
  multiphoton pumping, 32–34, 36
  population inversion, 20
  two-photon pumping, 19
High harmonic generation (HHG), 68
Houard, A., 45–73

I

Inert gases, 37, 39
Inversion-free amplification, 66–67
Ionized nitrogen
  amplification, 63–64

controversy, 65–68
inversion-free amplification, 66–67
natural population inversion, 65
post-ionization state coupling, 65–66
recollisional excitation, 69–72
rotational population inversion mechanism, 67
self-seeded lasing, 62, 63
superradiant behavior, 67–69
ultrafast gain buildup, 64

J
Jeffries, J.B., 15
Jing, C., 75–87

K
Kartashov, D., 46, 89–117

L
Laser field, 80–82
Laser filamentation
altitudes, 89
atmospheric constituents, 89
deep-UV pumping radiation, 91
electronic levels, 91
femtosecond, 124
high-power femtosecond laser pulse, 90
nitrogen, 111–115
optical excitation, 91
Raman scattering, 90
scattering regimes, 89
self-focusing, 125
in transparent media, 89
vibrational manifolds, 90
white-light continuum, 125
Levis, R.J., 121–137
Li, G., 75–87
Li, Z., 75–87
Liu, Y., 45–73

M
Malevich, P., 124
Microwaves, 111–115
Miles, R., 16, 19–42
Mirrorless, 20
Mitryukovskiy, S., 98
Molecular nitrogen, 91–102
in femtosecond filaments (*see* Femtosecond filaments)
in laser filaments, 111–115
physical mechanisms, 102–106
theory, 106–111
Molecular rotations
coherent coupling, 80–82
N_2^+ laser, 82–84
polarization state, 76
pump-probe experimental setup, 76
pump-probe scheme, 76
real-time observation, 77–80
rotational quantum states, 80–82
strong-field-ionization-induced N_2^+ lasers, 76
strong-field molecular physics, 75
temporal resolution, 76
ultrafast dynamics, 77–80
ultrashort intense laser pulses, 75
Multi-photon pumping, 32, 34, 36
Mysyrowicz, A., 45–73

N
Nd:YAG/dye laser system, 11
Nd:YAG laser, 34
Neutral nitrogen
ambient air, 51
amplification, 51, 52
backward emission, 49–50
circularly polarized femtosecond pumping, 48–52
modeling and numerical simulations, 58–60
population inversion, 52–55
spatial profile, 50
spectrum analysis, 49–50
time-resolved measurements, 55–58
Ni, J., 75–87
Nitrogen atoms, 7–9
Nitrogen ion
high-energy laser beam, 84
P-branch transition, 80
pressure, 83
Raman lines, 86
N_2^+ laser, 82–84

O
Odhner, J., 128
Odhner, J.H., 121–137
Optical parametric amplifier (OPA), 61
Optical parametric chirped pulse amplification (OPCPA), 93
Oxygen air lasing, 30
Oxygen atoms, 3–7

P
Photo-ionization, 14
Photo-multiplier (PMT) detector, 92, 98
Photomultiplier tube (PMT), 4
Picosecond laser filaments, 96, 116
Population inversion, 52–55
Post-ionization state coupling, 65–66

R
Radar REMPI technique, 28, 34
Radial profiles, 14
Radiation, 10
Raman scattering, 46
Raman spectroscopy
  anti-Stokes-shifted, 123
  applications, 121
  atmospheric gases, 123
  atmospheric temperature profiles, 123
  backward lasing schemes, 121
  chemical structure, 122
  coherent optical probing, 121
  condensed-phase materials, 124
  detection schemes, 122
  electric fields, 122
  filament-assisted, 126–135
  gas-phase molecules, 123
  gas-phase species, 122
  homonuclear diatomic molecule, 121
  molecular specificity, 122
  narrowband pulses, 124
  photomultiplier tube, 123
  probability, 121
  pump pulse, 124
  single-beam variant, 124
  trace concentration, 123
  two-photon excitation schemes, 121
  universal detection method, 123
  UV laser, 124
Recollisional excitation, 69–72
Remote sensing
  atmosphere, 123
  backward lasing schemes, 121
  experiments, 122
  investigation, 121
  laser-produced plasma channel, 124
  requirements, 126
  rotational revival transient, 135
  vibrational modes, 121
Romanov, D.A., 128
Rotational quantum states, 80–82
Rotational Raman Scattering, 84, 85
Rotational Raman spectroscopy, 135
Rotational spectroscopy, 135
Rotational wave packets, 77–80
Ro-vibrational transitions, 11

S
Seed pulse, 51, 52
Shanghai Institute of Optics and Fine Mechanics (SIOM), 47
Shneider, M., 89–117
Spatial mode profiles, 33
Spatial profile, 50
Spatial resolution, 14, 15
Standoff detection, 125, 126
Stimulated emission (SE), 3, 10–14
Stimulated Raman gain (SRG), 124
Stimulated Raman scattering (SRS), 123
Superfilamentation, 116
Super-fluorescence, 32
Superradiance, 46, 61–72
  ionized nitrogen molecules (*see* Ionized nitrogen)
Super-radiance, 32

T
Three-photon pumping, 20
Time-resolved measurements, 55–58
Two-photon-pumped stimulated emission
  backward-generated SE, 3
  carbon monoxide, 10–14
  crossed-beam setup, 14, 15
  de-excitation channel, 3
  emission wavelengths, 2
  energy-level diagram, 2
  excitation, 2
  laser-spectroscopic techniques, 1
  modeling, 1
  nitrogen atoms, 7–9
  optical techniques, 2
  oxygen atoms, 3–7
  pumped energy, 1
  simultaneous absorption, 2
  stimulated emission, 10–14
Two-photon pumping, 19

U
Ultraviolet (UV) pulses, 46

V
Vacuum ultraviolet (VUV), 1, 2

X
Xie, H., 75–87
Xu, H., 75–87

Y
Yao, J., 75–87

Z
Zeng, B., 75–87
Zhang, H., 75–87
Zhang, H.S., 77

MIX
Papier aus verantwortungsvollen Quellen
Paper from responsible sources
FSC® C105338

FSC
www.fsc.org

If you have any concerns about our products,
you can contact us on
ProductSafety@springernature.com

In case Publisher is established outside the EU,
the EU authorized representative is:
Springer Nature Customer Service Center GmbH
Europaplatz 3, 69115 Heidelberg, Germany

Printed by Libri Plureos GmbH
in Hamburg, Germany